International Technology Flows and the Technology Gap

Jan Monkiewicz, born 1949, is Associate Professor of Economics at Warsaw University and Chief Expert at the Ministry of Foreign Trade. Consultant for the United Nations Industrial Development Organization, the World Bank, and the United Nations Center on Transnational Corporations. Numerous books in Polish and English.

Jan Monkiewicz

International Technology Flows and the Technology Gap

The Experience of Eastern European Socialist
Countries in International Perspective

Routledge
Taylor & Francis Group

LONDON AND NEW YORK

First published 1989 by Westview Press

Published 2018 by Routledge
52 Vanderbilt Avenue, New York, NY 10017
2 Park Square, Milton Park, Abingdon, Oxon OX14 4RN

Routledge is an imprint of the Taylor & Francis Group, an informa business

Library of Congress Cataloging-in-Publication Data
Monkiewicz, Jan.
    International technology flows and the technology
gap.
    Bibliography: p.
    1. Technology transfer.   2. Council for Mutual
Economic assistance.   I. Title.
T174.3.M643   1988   338.9'26   88-27838

ISBN 13: 978-0-367-00357-9 (hbk)

ISBN 13: 978-0-367-15344-1 (pbk)

To Grażyna, my wife
and our children - Karolina and Marek

Contents

Introduction

International technology flows are a permanent component of the world-wide circulation of economic resources. Once incidental and sporadic, they became a massive and highly structured phenomenon in the post-war period. From a macroeconomic point of view, they may be viewed basically as a tool for rationalizing the international division of labor in science and technology and as a means for backward regions, countries and industries to catch up technologically. The latter is of particular interest for all developing countries as well as for the socialist countries of Eastern Europe, in short, CMEA.

The thrust of this book is to assess under what conditions and to what extent international technology transfer may contribute to closing the technology gap in overall industrial activity. A standard explicit or implicit contention in the available literature provides a generally positive view of foreign technology infusion. It appears, however, that this could at least be questioned.

This book argues that a successful catching up operation via foreign technology infusion presupposes the fulfilment of three conditions: a national environment basic, conducive to innovation, an international set up conducive to transfer and the proper intensity of technology inflow. It is argued that in the case of CMEA countries all three elements were largely absent. Therefore, in the end, their attempts at progress have been turned down.

This book is organized in four parts. Part one sets the analytical framework. Chapter 1 provides a review of the determinants of national technological performance, chapter 2 discusses the relationships between foreign technology inflows and national technological build-up and chapter 3 applies the concept of economic power to East-West technology transfer.

Part two attempts to quantify East-West technology flows in the 1970s and the beginning of the 1980s. To provide a proper background, chapter 4 begins with a brief presentation of the trends and patterns in international technology flows. Thereafter, principal characteristics of foreign technology infusion into CMEA countries are dealt with in chapter 5. Chapter 6, on the other hand, is devoted to the analysis of the role played by foreign investments.

Part three intends to analyse the dynamics of the CMEA technological position throughout the 1970s and the 1980s. It starts with a discussion of the concept of technology leads and lags and relevant measurement indicators (chapter 7). Thereafter, the evolution of the CMEA technological position in embryonic, potential and applied technology is discussed.

The fourth and last part provides principal findings and conclusions paying special attention to the role of technology transfer.

This book is based on vast empirical material collected during 1984-1987, both in Poland and the Federal Republic of Germany.

In preparing it I profited from the discussions with many people who shared their ideas, offered their advice and assistance. Among them I would like to mention particularly Prof. Jerzy Ruszkiewicz from the Warsaw Technical University, Dr A. Wass von Czege, Professors K.E. Schenk and S.G. Schoppe from the Hamburg University and Mr. Tom Ganiatsos from UNCTAD Technology Division.

I would like to express my special gratitude to the Alexander von Humboldt Stiftung which organized and financed my stay in the Federal Republic of Germany in 1985-1986 and thus helped to finalize the basic part of this book.

Jan Monkiewicz

Part I
Setting the Analytical Framework

Chapter 1
Determinants of National Technological Performance

1. Introduction

The subject of national technological performance and the resultant national technological build-up, i.e., accumulation of relevant resources, skills and capabilities, has since long been the matter of extensive studies of numerous researchers. They have undoubtedly cleared up a number of pertinent questions and eliminated some oversimplified views spread in the past. However, new questions and issues have been posed which await subsequent explanation. At the same time, some of the most challenging questions have remained untouched. These include, for example, the determinants of the rise and fall of individual countries and regions in the world-wide technological order, the role of leaders and followers in shaping world-wide technological trajectory, the national and international consequences of the assymetrical distribution of world-wide technological potentials and the like.

The aim of the present chapter is to provide a brief overview of the available theoretical explanations on determinants of national technological performance and comment on their explanatory power, i.e. their apparent ability of interpreting existing reality.

In the broad array of relevant literature, three principal methodological approaches have been identified and discussed. Let us call the first one a <u>transaction-specific approach</u>, the second one <u>institutional approach</u> and the third <u>historical approach</u>.

2. Transaction Specific Approach

The transaction specific approach towards national technological build-up and its determinants represents apparently the main stream of the relevant discussion so far. Obviously, it has evolved over time and became more and more refined. Its main methodological assumptions, however, remained basically unchanged.

According to this line of thinking, technological build-up is viewed as a specific process of the production and utilization of technological knowledge with different stages and actors. This process has its internal logic and is governed by relevant laws and interrelationships and its final results depend on the fulfilment of the necessary requirements, provision of the appropriate factors and on the accomplishment of their appropriate composition. The tradition for such an approach was laid down by the conference of the US Universities - National Bureau Committee for Economic Research and the Committee on Economic Growth of the Social Science Research Council held in 1962. The proceedings, published subsequently in the volume entitled "The Rate and Direction of Inventive Activitiy", have been considered a classical textbook on this subject (1).

The numerous contributors of the volume, guided apparently by a similar vision of technological change, drew attention to a variety of factors determining technological performance. Thus, for example, J.R. Minasian stressed the role of research and development outlays claiming that "productivity increases are associated with investment in the improvement of technology and the greater the expenditures for research and development the greater the rate of growth of productivity" (2). J. Schmookler, on the other hand, pointed out the role of interaction between accumulated knowledge and industry, assuming that the growth of modern western industrial technology has been primarily the result of an interplay of 1) changes in the state of knowledge and 2) changes in industry (3) (see Fig. 1.1.).

W.R. Thompson in the same volume drew attention to locational factors indicating that inventions were positively correlated with the rate of urbanization, scale of employment and degree of development of industrial and technological complexes (4).

The three authors just quoted are not meant to represent the major arguments contained in the analyzed volume but are rather to be taken as examples of the variety of factors studied and the variety of answers offered.

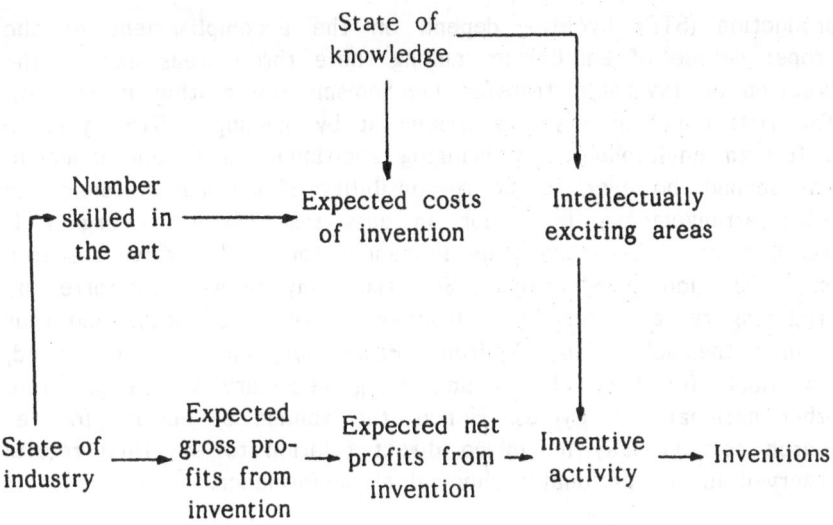

Fig. 1.1. Determinants of industrial inventive activity according to J. Schmookler

They all had in common, however, a perception of technological build-up as a process determined by some objective forces which should be sought in the very nature of technology creation and distribution. Paradoxically, however, the very process of technology creation and implementation has been largely considered the black box which only much later received proper attention and was extensively analyzed (5). Here one should mention particularly the works of C. Freeman, L.G. Soete, K. Pavitt, J.M. Utterback, Ch.T. Hill, F.M. Scherer, J. Röpke, G. Mensch and many, many others (6). The common perception of this process may be well exemplified by a model elaborated in the 1970s by a Polish economist, J. Ruszkiewicz (see Fig. 1.2.).

According to this model, technology build-up is viewed as a linear sequence of the creation of information (research), its transformation (development) and subsequent application in the production. The model draws attention to the fact that the elements research, transformation and production are both relatively interdependent and independent areas of human activity with different aims, value systems and applied languages. Final results of the process - frequently referred to as science-technology-production (STP) cycle - depend on the accomplishment of the proper degree of equilibrium among these three areas and on the creation of favorable transfer mechanisms (integrating interfaces). The first condition may be arrived at by opening a STP cycle to a foreign environment (by bringing additional supply and demand). The second, however, is the responsibility of national systemic and other arrangements. It is not an easy task because - argues J. Ruszkiewicz - there are some immanent forces that drive research and production activity apart. Scientists may be well comforted by producing research results exclusively for other scientists and thus feeding themselves ad infinitum. Production, on the other hand, may look for help abroad and bring necessary knowledge from other national STP cycles. Hence, the volume of outlays for research activity may be in no direct relation to the final results observed in the national technological performance.

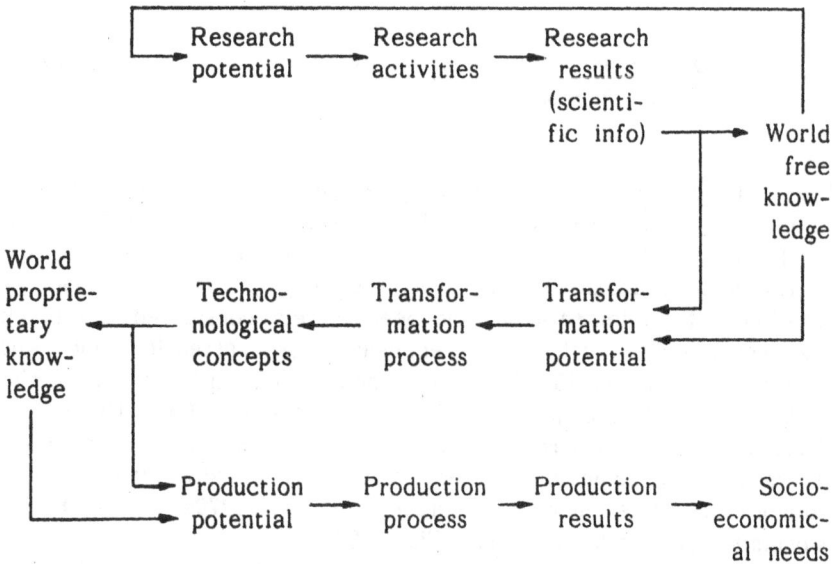

Fig. 1.2.   The process of technological build-up according to J. Ruszkiewicz

Source: Adapted from J. Ruszkiewicz - Problemy planowania i zarządzania w cyklu nauka-technika-produkcja, NOT, Warszawa 1977, p. 20

The model outlined above - a representative of the whole family of more or less similar conceptual constructions - is a typical static creature which permits us to understand how things run as soon as the relevant elements of an appropriate scientific-technical infrastructure have been created. It does not, however, explain how the relevant infrastructure is set up or, in other words, what steps are to be made on the way to the accomplishment of a national technological capability and what are the basic levels of this capability.

These questions gained particular importance in view of the needs of the developing countries plagued by severe technological backwardness.

It is no doubt therefore that they have been studied mainly by researchers working on industrialization processes in developing countries (7). Apparently one of the most interesting contributions has been offered by L.E. Westphal (8) and a group of other researchers cooperating with him in the course of the broad empirical research program launched by the World Bank at the end of the 1970s.

According to this approach, national technological capability, understood as "the ability to make effective use of technological knowledge", may be divided into three broad areas: production, investment and innovation (9). Technological capability is not acquired by a one shot operation. It is rather a painful and time consuming process that adds little by little to the overall technological development of a nation. Production capability comprises the knowledge and skills which are necessary for the successful operation of existing plants and production lines. It consists of a production management capacity, production engineering, repair, maintenance and marketing. Investment capability is composed of three interrelated components: 1. project preparation capability, comprising identification and selection of the technology which is best suitable for the local environment, 2. project execution capability, encompassing the knowledge and skills for the execution of the given development projects, and 3. the capital goods capability, i.e. the ability of manufacturing new machinery and equipment (see Table 1.1).

The third component of technological capability is an innovation capability, reflected in the nation's ability to create new products and processes and diffuse them throughout economic activity.

Table 1.1. Elements of Production, Investment and Innovation Capability

Production Capability (a)

Production management - to oversee the operation of established facilities

Production engineering (b) - to provide the information required to optimize the operation of established facilities, including:
* Raw material control - to sort and grade inputs, seek improved inputs
* Production scheduling - to coordinate production processes across products and facilities
* Quality control - to monitor conformance with product standards and to upgrade them
* Trouble-shooting - to overcome problems encountered in the course of operation
* Adaptations of processes and products - to respond to changing circumstances and to increase productivity

Repair and maintenance of physical capital - according to regular schedule or when needed

Marketing - to find and develop uses for possible outputs and to channel outputs to markets

Investment Capability

Manpower training - to impart skills and abilities of all kinds

Preinvestment feasibility studies - to identify possible projects and to ascertain prospects for viability under alternative design concepts

Project execution - to establish or expand facilities, including:
* Project management - to organize and oversee the activities involved in project execution
* Project engineering - to provide the information needed to make technology operational in a particular setting, including:
  - Detailed studies - to make tentative choices among design alternatives
  - Basic engineering - to supply the core technology in terms of process flows, material and energy balances, specifications of principal equipment, plant layout

- Detailed engineering - to supply the peripheral technology in terms of complete specifications for all physical capital, architectural and engineering plans, construction and equipment installation specifications
- Procurement - to choose, coordinate, and supervise hardware suppliers and construction contractors
- Embodiment in physical capital - to accomplish site preparation, construction, plant erection, manufacture of machinery and equipment
- Start-up of operation - to attain predetermined norms

Innovation Capability

Creative capacity - generation of new original technical ideas of practical utility

Research and development management including:
* selection of research projects
* execution of research projects
* adjustment of technical ideas to practice

Innovation management - to ensure a flow of inventions to production sphere including:
* implementation activities
* diffusion of inventions throughout economic activities
* utilization of industrial property rights to secure additional economic benefits

---

(a) The activities listed refer to the operation of manufacturing plants, but similar activities pertain to the operation of other types of production facilities as well.

(b) The term "production engineering" departs from conventional usage in that it is far more broadly defined to include all of the engineering activities related to the operation of existing facilities. Here the term encompasses "product design" and "manufacturing engineering" as these terms are generally used in reference to industrial production. See the entries under these headings in the <u>McGraw-Hill Encyclopedia of Science and Technology</u> (New York: McGraw-Hill Book Company, 1977).

Source: Production and investment capabilities are taken from L.E. Westphal, L. Kim, C.J. Dahlman - Reflections on Korea's ... op.cit., p. 7; innovation capabilities are added by the author

National technological development proceeds from the simplest levels of production capability, through its advanced stages to investment capability and thereafter to innovation capability. The velocity of movement along this path is basically attributed to the so-called environment for technological change. Thus, as is clear from our presentation, the basic proposition of the model lies in reversing the standard view (well suited for industrialized countries) of technological change from a sequence of innovation to investment to production, a reverse to sequence; production to investment to innovation. In suggesting this, it is argued that "innovation often is the product, however, of experience in other investment and production activities. And the activities in new investment and production often lead to minor innovations" (10). Possibilities for reversing the sequence described above result from the fact that developing countries may in many instances acquire the necessary technology abroad and thus use subsequent production capability as "the foundation for developing capabilities in investment and innovation" (11).

A special mutation of the transaction specific approach is associated with studies identifying some secular trends in technological activity, i.e. theories which explain varying technological performance over time by pointing out the existance of some long run patterns (12). Apparently the best example of such an approach is a theory of technological stalemate offered by G. Mensch in his famous book published in the mid-1970s.

The theory departs from the observation, provided earlier by Kondratiev and Kuznets, that the economic development of nations in an industrial era is characterized by some long cyclical swings lasting on average around 50 or so years (13). These long swings, accordingly, are interpreted as an outcome of a changing interplay between innovations and economic activity. In analyzing this interplay, G. Mensch distinguishes between so-called basic innovations and improvements. The first radically change existing reality and provide for the emergence of new trade and industrial branches whereas the second simply extend these basic innovations and contribute to their further development and improvement (14). Basic innovations find their way into production only during deep recessions when it is no longer possible to rely on improvements and thus provide an impetus for the subsequent cycle which lasts again as long as major opportunities offered by them are utilized by a wave of improvements. The major reason behind such a phenomenon is a conservative behavior of established producers and

22

branches reinforced by governmental protectionist policies. To prove his hypothesis, G. Mensch points to the fact that basic innovations always come in waves and always follow deep depressions (see Fig. 1.3.).

m_i = frequency of basic innovations in 1740–1960

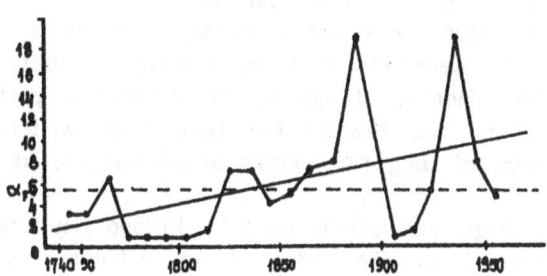

Fig. 1.3. Long waves in innovative activity (15)

Thus he formulates, in contrast to previous implicit assumptions, a principle of discontinuity in technological change. Incidentally, he also seems to indicate that in the long run there is an upward trend in the number of basic innovations appearing on the market and that swings (discontinuity) become ever larger.

Thus, according to this theory, the varying technological performance of countries over a period is predetermined by their actual location on the long technological cycle. It will boom in times of basic innovations and slack stepwise in the following years, before a new upheaval in basic innovations occurs. An interesting aspect of this contention, which has been disregarded by Mensch, would be the question whether these waves have a global nature and whether one could think of regionally or nationally determined long technological waves and, thus, what would be the consequences of such a disintegration of basic/improvement innovation cycles for international technological set up.

The critical element of G. Mensch's theory, on which the subsequent statistical analysis has been based, is undoubtedly the concept of basic innovations. In identifying them, he relied primarily on secondary sources (16). The importance of this issue may be judged from a similar investigation carried out by a Russian economist, E.E. Filippovskij.

Like G. Mensch, E. Filippovskij concentrated his attention on large technical innovations and tried to investigate their dynamics in the US economy during the time span 1921 to 1967. His very extensive statistical calculations and analyses were based on the data provided for US Censuses of Manufacturing of 1921, 1937, 1947, 1958 and 1967. Hence, altogether, he covered a period of approximately 50 years, i.e. exactly the time span sufficient to discover the existence of long run swings in innovations, as stipulated by Mensch.

The concept of large innovations used by Filippovskij was very simple - he considered as such only those technical solutions, which have subsequently born new branches of industrial activity and thus have been reflected in the new entries in the Census of Manufacturing (17). These entries are made only when new production attains sizeable volume.

In his analysis, Filippovskij identified 1116 new entries in the period covered by available data and he found them to be nearly equally distributed over the time span. Therefore, he concluded by pointing out that the trend of large innovations in the last 50 or so years was constant and no traces of dramatic shifts could be statistically proved (18).

The transaction specific approach roughly sketched above has contributed substantially to our understanding of national technological performance. It drew attention to the fact that the innovation process requires specific resources and specific conditions, and it is also governed by some specific laws and regularities. By discovering and investigating these one could thus get things under control.

What this approach fails to recognize and explain are instances in which apparently the same innovative effort brings extremely different results, what determines the scope of national innovative effort itself, how is an innovative conducive environment created, etc.

It is basically both static (i.e. it principally analyzes technological build-up within certain existing conditions) and fragmentary (i.e. in most cases only selective variables are taken into account).

This approach is not, in fact, directly involved in inter-country comparisons in national technological performance, and hence, is not looking for national differences in this area but rather tries to detect and pull together commonalities.

This is precisely where the institutional approach comes to the forefront as it is more concerned with international differences than commonalities.

3. Institutional Approach

The essence of the institutional approach lies in associating national technological performance with the characteristic features of national economic organization. It is claimed, explicitly or implicitly, that to understand varying national performance in technological build-up, systemic variables must come to the forefront of the analysis. Thus, in contrast to the transaction specific approach, the central objects of analysis are no longer specific attributes and requirements of technological build-up and the degree of their satisfaction, but the organizational set up in broad terms, in which this process has to take place.

The origins of this approach are closely bound to Marxian economic writings and particularly to Marx's analysis of capitalism. As is well known, the central variable of the Marxian theory of economic development is that of the nature of the ownership of the means of production, which subsequently determines the socio-economic organization of the production processes and has decisive bearing on the economic performance of a given society. Marx's position in this respect is well illustrated by the following quotation from his Communist Manifesto: "... the bourgeoisie, during its rule of scarce one hundred years, has created more massive and more colossal productive forces than have all preceeding generations together" and further on: "The bourgeoisie cannot exist without constantly revolutionizing the instruments of production, and thereby the relations of production, and with them the whole relations in society. Conservatives of the old modes of production in unaltered form, was, on the contrary, the first condition of existence of all earlier industrial classes" (19).

The Marxian argumentation was basically repeated subsequently by J. Schumpeter in his analysis of the economic workings of capitalism (20). The new elements brought forward by him were his observations regarding socialized economic systems, which he basically viewed as the ones which, by eliminating entrepreneurial functions (and thus profit-seeking as the principal motive behind technological change), would be characterized by slow and routinized technological change.

A somewhat different perception was offered by F. Hayek who, in contrast to Schumpeter, concentrated on the rationality of the resource allocation and claimed that capitalist market arrangements are a better device for providing a rational allocation of resources than socialist central planning. He principally attributed the reasons for this to the underlying information processes (21).

These two principal suppositions, i.e. the role of motivation and information, have for a long time been the guiding principles of the institutional approach to national technological performance and have been clearly dominated by East-West comparisons without serious attempts of intra-systemic examinations.

A fresh impetus for institutional thinking has been provided by the so-called new institutional economy, largely associated with the works of R. Coase and O. Williamson and particularly by the so-called theorem of institutional choice developed by K.-E. Schenk (22). Whilst R. Coase and O. Williamson restrict themselves primarily to micro-economic considerations, K. Schenk upgrades the analysis to encompass both the meso and macro-economic levels.

According to this line of thinking, the economic performance of individual agents could best be explained by referring transaction-specific requirements to the coordinating mechanis in use, of which pure market and pure hierarchical relations constitute two extreme forms of possible coordination arrangements. Thus, there is a need for analyzing and explaining pertinent characteristics of individual types of transactions and attributes of different institutional arrangements.

According to O. Williamson, there are three principal characteristics in each transaction:

a) the degree of uncertainty, as far as its realization and results are concerned,
b) the time density of a transaction (its frequency) and its duration,

c) the degree of its idiosyncrasy in terms of the required human and fixed capital (23).

The transactions may take place in a continuum of institutional arrangements, spanning from pure market to pure hierarchy, the principal characteristics of which are the relevant coordination mechanisms: in the case of pure market, these are horizontal relations among independent actors, in the case of hierarchy, these are vertical relations among subordinated and controlling actors (24). What is important for the latter is at what level of the vertical structure the transactional decisions are taken (location of a transaction-potential) and at what level the relevant information is available (location of information-potential) (25).

On the basis of transaction-specific characteristics and coordination-specific features, one can identify the optimal coordination mechanisms required (see Fig. 1.4.).

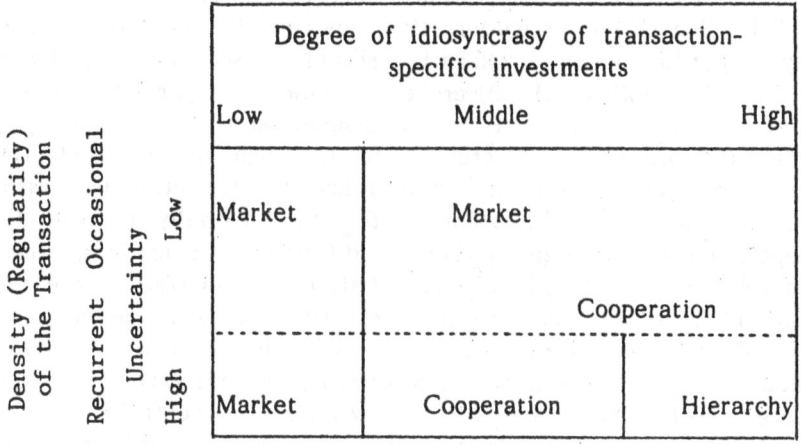

Fig. 1.4. Nature of transaction and required coordination mechanism

Source: J. Harders - Arzneimittelforschung und Industrieorganisation DDR und Ungarn im Vergleich, Ökonomische Studien, Band 37, Stuttgart, New York 1985

To apply this general model to technological build-up, a list of principal characteristic features of this "transaction" has to be elaborated. Investigating this issue, J. Harders comes to the conclusion that the pertinent features of technological build-up transactions (innovation transaction) may be described in the following way:

1) high degree of uncertainty,
2) low recurrency (frequency) of the transactions,
3) high degree of transaction-specific investments (26).

Thus, it turns out that the most suitable coordination mechanisms required are some forms of cooperation arrangements and/or hierarchical schemes (27). This contention is, on one hand, in apparent agreement with the prevailing internalization of R+D activities observed in the companies of the industrialized West. On the other hand, however, it is in obvious disagreement with most of the literature on the centrally planned economies (hierarchical systems) which claims that the poor technological performance of these countries results precisely from their institutional arrangements, based predominantly on hierarchical mechanisms (28).

K.E. Schenk, investigating this question in a more detailed way, suggests that an explanation should be sought principally in the varying ability of different systemic arrangements in risk management. He departs from the assumption that innovation processes are primarily characterized by the high risk involved. This risk is reduced by providing more relevant information (scientific, technological, economic, commercial) which normally requires the cooperation of a large number of different organizations, whose composition is constantly changing (different organizations are relevant in different phases of the innovation process). The coordination of their activities, however, is difficult to attain in hierarchically organized systems. Additionally, in hierarchical system (more precisely, in hierarchical systems of centrally planned economies) an assymmetrical distribution of the information and transaction potential exists, which aggravates the relevant problems (see Fig. 1.5.) (29). As a result, the decision making level is underinformed, whereas information holders are deprived of the rights for decision making. In effect, the degree of innovativeness of the hierarchical systems is relatively reduced. K. Schenk also brings to attention the fact that in the hierarchical systems of the socialist countries, motivation instruments are largely production target oriented, which adds to their limited propensity to innovate.

Market-type coordination
(West)

Hierarchical coordination
(Socialist countries)

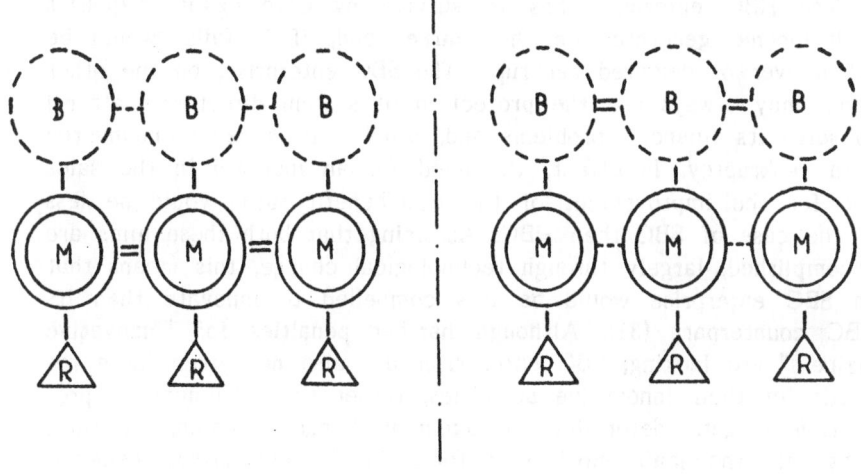

Notes:

⊚ Principal, innovation-specific information potentials

(⁻) Less relevant information potential

═ Principal, innovation-specific transaction potentials

---- Less relevant transaction potentials

B - bureaucracy; M - management; R - realization level

Fig. 1.5. Distribution of information and transaction potentials in innovation processes in market type and hierarchical type system according to K.-E. Schenk

An alternative explanation on the impact of different systemic arrangements on innovation processes has been offered by a Hungarian economist, J. Kornai (30).

He describes two different systemic arrangements which have decisive bearing on the behavior of the relevant economic agents:

29

one in which enterprises are subject to hard budget constraints (HBC) and the other in which they operate under soft budget constraints (SBC). The first are dominant in market arrangements, the second in hierarchical systems.

The HBC enterprise has to survive by covering its expenses with income generated on the market and, if it fails, would be taken over or declared bankrupt. The SBC enterprise, on the other hand, may always use the protection of system directors (patrons) to solve its financial problems and, usually, it is never endangered with bankruptcy. In effect, the need for an increase in the sales and for the improvement of the cost/benefit ratio would be less in the case of SBC than HBC. Assuming that both these aims are accomplished, largely through technological change, this means that an SBC enterprise would be less compelled to innovate than its HBC counterpart (31). Although harsher penalties for "innovative laziness" are lacking, SBC enterprises are also not given large rewards for their innovative activities, rather the fulfilment of production targets determine management bonuses, which, in turn, constitute the main motivating force in the enterprise's behavior (32). An increase in the production level due to technological change would bring only limited financial satisfaction for the management concerned, as:

a) it has to be weighed against the losses incurred in bonuses at a time of innovation implementation,

b) it would be quickly swallowed up by the planning through so-called "rachet effects", i.e. higher production targets would be set up for the enterprise in the subsequent period to allow for its increased production capacity.

To summarize, SBC enterprises would exhibit a lower innovative propensity than their HBC counterparts. Thus, departing from different assumptions, Kornai's SBC/HBC theory also provides different interpretations of the inter-systemic differences in technological performance than those offered by K.-E. Schenk. While both of them believe in the risk-reward syndrome, Schenk emphasizes high uncertainty of innovation activities and the need both for actions to ameliorate it and for stimulis to reward it (which according to him are apparently less well devised in hierarchies), Kornai stresses the lack of negative consequences (penalties) for not innovating (innovative laziness) and also the general lack of motivation to promote innovations at the enterprise level. In other words, for Schenk the prevailing situation is characterized by a high risk - low reward relationship, whereas for Kornai, low risk is coupled with an even lower reward.

As can be seen from the majority of the preceding analysis, institutional thinkers favour markets over hierarchies and, as far as technological performance is concerned, point out that hierarchical systems have far more drawbacks than market arrangements for providing successful innovations. They rarely try to weigh these drawbacks against the merits of the hierarchical (i.e. planned) systems. One of the positive exceptions in this respect is a recent study by P. Hanson, in which a more balanced view is offered (33).

Summarizing his insightful analysis on the impact of institutional structures existing in centrally planned economies upon technological performance, Hanson concludes that their apparent strengths are related to:

1) the capacity to generate a relatively high rate of investment, both in human as well as in physical capital,
2) the capacity to design and implement industrial policies which take account of the externalities so important in the case of innovation activity,
3) the capacity to eliminate unnecessary duplication of R+D activities.

Their weaknesses, on the other hand, are related principally to the lack of properly structured motivations for innovative enterprises (34).

Assessing the institutional interpretation of varying technological performance in different nations, it seems fair enough to conclude that it undoubtedly broadens and deepens our understanding of national technological performance by adding a new perspective to it. At the same time, however, in its present form, it has several weaknesses.

One of these is the predominant concentration on the microlevel to explain macro-processes and/or the utilization of microlevel models for describing macro-relevant phenomena. This has both some advantages as well as disadvantages. In the first instance, it enables a better understanding of the enterprise behavior, its determinants and its impact on macro-processes. At the same time, however, it may be misleading when used for explaining macro-processes alone. Thus, for example, a distinction of HBC and SBC enterprises is very helpful indeed to understand their varying behavior and their different attitude towards innovations. At the same time, however, it says little about the innovation propensity of the entire system (i.e. national economy) which is not simply a sum of enterprises. Furthermore, the entire system is in the end HBC, but does this mean that it will behave in the manner ascribed to HBC enterprises, or perhaps otherwise? This concentration on micro-aspects in explaning the technological performance of individual systems reflects undoubtedly the role played by enterprises in market type arrangements. They are principal actors of the game as well as major economic decision makers. However, if one looks closely at hierarchies then one can immediately see that above the major actors and major decision makers there are various levels of bureaucracy. If so, then one should logically draw ones attention to their behavioral characteristics and relevant determinants. Thus, for example, in Schenk's discussion of the innovation risk in the market system, the risk is with the innovating enterprise and in hierarchical systems it is with the hierarchy. Now, the questions are whether it is perceived at all, what forms does it take, what are the consequences, etc., etc.

What is suggested here is that institutionalist thinkers apparently utilize straightforwardly some schemes pertinent to market systems in a different context of the hierarchical systems, and, hence, frequently waste their efforts on the analysis of secondary elements.

Three other features of the current institutional approach deserve critical assessment as they seem to distort the soundness of the subsequent analysis. The first one is the implicit belief that institutional arrangements, whatever they may be, will yield the same consequences wherever they are applied, i.e. the apparent neglect of socio-cultural and development variables. Economic history provides many examples that this is a severe simplification. Both the same market as well as hierarchical schemes may apparently bring different results, depending on the socio-cultural and developmental context of its utilization.

The second one is the frequent confusion of systemic and policy related variables. The question to be answered is what characteristics are derived from the logic of the given systemic arrangements and what are the results of the pursuit of a given socio-economic policy. If we come again to the question of the innovation rewards offered by hierarchies, for example, the problem could be formulated in the following way - is a production target orientation of reward schemes in hierarchical settings a systemic data or a matter of policy priorities?

Lastly, one should point out that although the current stream of institutional analysis recognizes the fact that there is a broad array of market economic systems within which "there may be different market structures and property arrangements which have radically different implications for innovative performance", it seems to be very conservative as far as hierarchical systems are concerned (by principally taking the one extreme Soviet type form) and rejects basically the same rights for them (35).

4. Historical Approach

The essential feature of the historical approach is the interpretation of national technological performance in terms of the variables pertinent to the time and place for which the analysis is carried out. Thus, in contrast to the two former methodological approaches which were oriented towards identifying some universal and secular factors (in transaction-specific literature - relating to the process of technological change per se; in institutional analysis - relating to its macro-organization), the historical approach is rather concerned with the sets of specific factors and forces in operation.

According to this line of thinking, an assessment of the national technological performance requires at the onset an analysis of the technological level achieved so far. It is asserted, that technological build-up is largely a cumulative and self-generating process. Nathan Rosenberg, a leading representative of the historical approach, speaks in this respect about the compulsive sequences and cites many historical occurances confirming this hypothesis (36). The technical imbalances behind compulsive sequences led, for example, to the invention of Bessemer's process, major breakthroughs in the machine tool industry in the 19th century or to the development of armored ships in response to the increase in the destructive power of offensive weapons (37).

The next issue raised by the historical approach is the impact of rapid changes in the conditions for economic activities and, in particular, the disruption of the accustomed sources of supply, in most cases provoked by the outbreak of wars (38). Many examples could be cited in favor of this contention. Thus, for example,

France's leadership in the synthetic alkali industry emerged as a by-product of the Napoleonic Wars and resulting in a disruption of the supply of Spanish barilla; or the cutting off of American supplies to Britain during the American Civil War which served as an important inducement to technical change in British textiles. The Japanese occupation of the South-East Asian natural rubber producers culminated in the emergence of the US synthetic rubber industry (39).

An important role among the factors determining national technological performance is assigned to cultural values and social structures. In his illuminating discussion of the American technological performance in the 18th and 19th century, N. Rosenberg pointed out that apparently it was the cultural and social values of the American society which permitted the United States to take a world lead in the modern industrial economy. What is interesting is that it happened at a time when the US was at least half a century behind the UK in its industrial development (40).

A mass-production technology, based on the progressive assembly of interchangeable parts, produced by a sequence of highly specialized machines (a characteristic feature of American technological development) could not have been born in Britain because - as N. Rosenberg argues - British producers were under much greater pressure to customize their products. Thus, it is not by coincidence that both mass gun and car manufacturing started in the relatively backward US and not in the United Kingdom - at that time world technological leader.

The same socio-cultural characteristics apparently explain much of the Japanese technological performance as well as South Korea's recent advancement (41). Equally important for national technological performance are economic characteristics, including such dimensions as industrial structure, presence or absence of raw materials, size of labor force and its qualitative parameters, distribution of economic potential by firms (the relative share of big and small business), etc. (42)

The historical approach pays also a great deal of attention to the policy devised variables, i.e. the ability of individual countries to design proper technological and industrial policies (43).

5. Concluding Remarks

The foregoing discussion, though very concise and obviously in many respects quite superficial, demonstrates beyond a doubt that an explanation of national technological performance cannot be provided successfully at a highly abstract level by concentration on one or few systemic or transaction oriented characteristics. At the same time, it is equally fruitless to concentrate exclusively on some local determinants and characteristics insofar as some more general conclusions are to be offered.

It seems reasonable to conclude that what we need is an interdisciplinary and eclectic approach that would utilize the propositions contained in all three aforementioned groups of methodological approaches. What is important, is to accomplish a proper equilibrium between the Scylla of generalities, as advanced by transaction type and institutional type models, and the Charybdis of specificity and uniqueness, as advocated by historical explanations.

References and Footnotes

1   The rate and direction of inventive activity, Princeton University Press, Princeton, 1962
2   J.R. Minasian - The economics of research and development, in: The rate and direction ... op.cit., p. 94
3   J. Schmookler - Changes in industry and in the state of knowledge as determinants of industrial invention, in: The rate and direction ... op.cit., p. 195
4   W.R. Thompson - Location differences in inventive effort and their determinants, in: The rate and direction ... op.cit., pp. 253-272
5   For a review of the concepts, see O. Granstrand, J. Sigurdson (eds) - Technological and industrial policy in China and Europe, Technology and Culture, Occasional Report Series no. 3, Research Policy Institute, Lund 1981, pp. 8-15
6   Ch.T. Hill, J.M. Utterback - Technological innovation for a dynamic economy, Pergamonn Press 1979; F.M. Scherer - Innovation and growth, Cambridge, Mass., MIT Press 1984; J. Röpke - Die Strategie der Innovation, Tübingen, Mohr 1977; G. Mensch - Das technologische Patt, Umschau Verlag, Frank-

furt am Main, 1975; C. Freeman - The economics of industrial innovation, Penguin Books, 1974; K. Pavitt, L. Soete - International differences in economic growth and the international location of innovations, SPRU, Brighton, May 1981

7   See special issue of The Annals of the American Academy of Political and Social Science, November 1981 which was fully devoted to these problems

8   L.E. Westphal, L. Kim, J.C. Dahlman - Reflections on Korea's acquisition of technological capability, The World Bank, April 1984, Discussion paper, Report no. 77, p. 5; C.J. Dahlman, B. Ross-Larson, L.E. Westphal - Managing technological development. Lessons from the newly industrializing countries, World Bank Staff Working Papers no. 717, The World Bank 1985

9   L.E. Westphal, L. Kim, J.C. Dahlman - Reflections on Korea's acquisition ... op.cit., p. 5

10 C.J. Dahlmann, B. Ross-Larson, L.E. Westphal - Managing technological development ... op.cit., p. 18

11 C.J. Dahlmann, B. Ross-Larson, L.E. Westphal - Managing technological development ... op.cit., p. 18

12 For an illuminating discussion of these theories, see Ch. Freeman - Innovation as an engine of economic growth: retrospect and prospect, in: H. Giersch (ed.) - Emerging technologies: consequences for economic growth, structural change and employment, J.C.B. Mohr, Tübingen, 1981, pp. 1-26

13 N.D. Kondratiev - The major economic cycles, Lloyds Bank Review, no. 129, London 1978, pp. 41-60

14 G. Mensch - Zur Dynamik des technischen Fortschritts, in: Zeitschrift für Betriebswirtschaft, 41/1971, pp. 295-314

15 G. Mensch - Das technologische Patt. Umschau Verlag, Frankfurt 1975, p. 142

16 G. Mensch - Das technologische Patt. Umschau Verlag, Frankfurt 1975, p. 133-134

17 E.E. Filippovskij - Obnovlenje promyshlennoj produkcji w kapitalisticheskikh strankah, Izd. Nauka, Moskwa, 1978, pp. 35-39

18 E.E. Filippovskij - Obnovlenje ... op.cit., p. 86

19 K. Marx, F. Engels - The Communist Manifesto, 1951, pp. 36-37

20 J. Schumpeter - Capitalism, Socialism and Democracy, New York, 1950

21 F. Hayek - The use of knowledge in society, American Economic Review, vol. 35, no. 4, pp. 519-530

22 R. Coase - The nature of the firm, Economica, no. 4/1937, pp. 386-405; O. Williamson - Markets and hierarchies: analysis and antitrust implications, New York, London 1975; K.-E. Schenk (ed.) - Vergleichende System- und Industriestudien - Ein "Institutional Choice"-Ansatz, Stuttgart, New York, 1983, Ökonomische Studien, Band 34

23 O. Williamson - Transaction-Cost-Economics; The governance of contractual relations, Journal of Law and Economics vol. 22, 1979, pp. 233-262

24 K.-E. Schenk - Märkte, Hierarchien und Wettbewerb - Elemente einer Theorie der Wirtschaftsordnung, München 1981

25 K.-E. Schenk - Zum Vergleich der Innovationsorganisation in unterschiedlichen Wirtschaftssystemen, K.-E. Schenk (ed.) - Studien zur politischen Ökonomie, Ökonomische Studien, Band 32, Stuttgart, New York, 1982, pp. 85-97

26 J. Harders - Arzneimittelforschung und Industrieorganisation ... op.cit., pp. 17-20

27 J. Harders - Arzneimittelforschung ... op.cit., p. 45

28 J.S. Berliner - The innovation decision in Soviet industry, Cambridge, MIT Press, 1976; A. McAuckley - The incompatibility of central planning and rapid innovation, mimeo, University of Essex, 1984; R. Amann, J.M. Cooper (eds.) - Industrial innovation in the Soviet Union, New Haven, Yale U.P., 1977; P. Hanson, K. Pavitt - The comparative economics of research, development and innovation in East and West: a survey, SPRU, University of Sussex, February 1987

29 K.-E. Schenk - Zum Vergleich der Innovationsorganisation ... op.cit., pp. 85-97

30 J. Kornai - The economics of shortage, Amsterdam, North Holland, 1980

31 An extensive analysis of these relationships is offerend by P. Hanson - The comparative economics of research, development and innovation: a survey, mimeo, CREES, University of Birmingham, undated

32 P. Hanson - The comparative ... op.cit., p. 13

33 P. Hanson - The comparative ... op.cit.

34 P. Hanson - The comparative ... op.cit., p. 19

35 P. Hanson - The comparative ... op.cit., p. 4

36 N. Rosenberg - Perspectives on technology, Cambridge University Press, Cambridge, New York, Melbourne, 1976, pp. 110-117

37 N. Rosenberg - Perspectives on technology ... op.cit., pp. 115-116
38 N. Rosenberg - Perspectives on technology ... op.cit., p. 121
39 N. Rosenberg - Perspectives on technology ... op.cit., p. 121
40 N. Rosenberg - Perspectives on technology ... op.cit., p. 286
41 J. Sigurdson - Japan's high technology race. The information technologies, Research Policy Institute, Technology and Culture, Occasional Report Series no. 8, Lund 1983
42 The conditions for success in technological innovation, OECD, Paris 1971, pp. 143-148
43 O. Granstrand, J. Sigurdson (eds.) - Technological and industrial policy in China and Europe, Technology and Culture, Occasional Report Series no. 3, Research Policy Institute, Lund 1981

Chapter 2
Foreign Technology and National Technological Build-up

A Search For a Framework

1. Introduction

When discussing determinants of national technological performance in the preceding chapter, no reference was made to the possible role of foreign technology. This issue comes to the center of our analysis is in the present chapter. In doing so, we depart from the observation that in modern times the technological development of nations has never been entirely a result of an internal generation of innovations. On the contrary, most countries acquired much of their technology from abroad. This phenomenon increased particularly with the rise of capitalism and the concurrent industrialization processes and it is even more pronounced today.

The question of technology transfer is a major theme in the North-South development dialogue as well as the East-West relations debate. It also seems to gain in importance with respect to intra-North technological flows, parallel to the rise in US technological protectionism.

Having said all this, and when one looks at the relevant economic literature, one could note with surprise how little attention has been paid so far to the theoretical framework explaining the basic relationship between the importation of technology and the national technological build-up, determinants of the international division of labor in science and technology, factors shaping the industrial and geographic structure of international technology flows, effects of technology exports and the like.

The present chapter attempts to elaborate on the first question, which is of utmost importance for technological laggards who are seeking to decrease their backwardness using, inter alia, a direct importation of foreign developed technology.

In the literature dealing with this subject, three principal groups of views can be identified. One, viewing the import of technology as an essentially positive factor of national technological build-up, the other, considering it as a developmental trap, and yet another, seeing it as a life preserver for some economies.

Let us call the first line of thinking the <u>production of development approach</u>, the next the <u>reproduction of dependency approach</u> and the third the <u>survival hypothesis approach.</u>

2. Production of Development Approach

A principal feature of this approach is a generally positive stance with respect to the application of foreign technology to national technological development, implying its positive impact on both national technological buildup as well as the resultant national economic progress. This kind of view was particularly spread wide during the 1960s, in the literature dealing with the development prospects of underdeveloped countries. It was by and large a straightforward extension of the idea that the technological build-up of a country depends on a supply of industrial innovations which, in turn, results from the respective innovative effort, in terms of the means allocated to this area of human activity. This reasoning, prevailing in the economic literature on technological change in the 1960s, within what we called a transaction specific approach, was applied in an unqualified manner to the imported technology. As commented by a leading British specialist on economics of technological change Ch. Cooper: "Hardly anyone asked how all-round dependence on technologies from the advanced countries affected economic and social development. Afer all, the underdevelopment of science and technology could be regarded as simply a particular aspect of the general condition of underdevelopment. And if, in the interests of development, it was necessary to strengthen scientific and technological capability, the thing could be done quite easily: all that was required was to endow a few good research institutes, and to train the appropriate number of scientists to do the research. <u>Moreover, the availability of production technologies from the industrialized countries was more or less an unqualified blessing</u> - less unqualified perhaps for development economists who made ritual noises about incompatibilities between factor endowments and capital intensity. Whatever problems there might be in the transfer of technology (and in the sixties' it was usually argued that the main problem was that there wasn't enough of it), it seemed that the underdeveloped countries could make great jumps by using the technologies which had already been developed in the advanced countries" (1) (underlined by the author). This mood of naive optimism was, inter alia, largely reflected in the practical activities of the UN system, which until the late 1970s clearly concentrated its effort on technology import and only thereafter shifted its attention more to the question of technological build-up (2).

The theoretical framework for this approach was provided by a proposition put forward already at the beginning of the 20th century by T. Veblen and further developed by A. Gerschenkron, and thus frequently referred to as the Veblen-Gerschenkron effect (3). Its essence could be best described by the following quotation from Gerschenkron: "Assuming an adequate endowment of usable resources, and assuming that the great blocks to industrialization had been removed, the opportunities inherent in industrialization may be said to vary directly with the backwardness of the country. Industrialization always seemed the more promising the greater the backlog of technological innovations which the backward country could take over from the more advanced country" (4) (underlined by the author).

This principal proposition has been further developed and refined by S. Gomulka, R. Findlay, J. Cornwall, H.W. Singer and L. Reynolds. In his original model, elaborated in 1970, Gomulka argues that the effects of technology import on the technological build-up of the recipient nation and the resultant economic growth will depend primarily on the size of the transfer flows and the size of the relevant technology gap vis-à-vis the supplying nations (5).

Assuming that the labor productivity in a given economy is a good proxy measure of the aggregate technological level, he spelled out the following principal relationships existing between imported technology and national technological build-up (6)

$$y_t = a_t + d_t \tag{1}$$

$$d_t = \frac{I_{t-l}}{K_{t-l}} \sum_i r_i \left(\frac{Y/L_{it-l}}{Y/L_{1t-l}} - 1 \right) \tag{2}$$

where:

Y   denotes the growth rate of the labor productivity of a technology receiving nation

t   denotes year

a   denotes labor producitivity growth attributed to indigenous technical change and to disembodied technology import

d   denotes diffusional effect (i.e. productivity growth due to embodied diffusion from abroad)

I   denotes machinery investment in the recipient country

K   is the importer's stock of machinery

l   is the lead time between the production of machinery and its installation (technological vintage)

r_i denotes the proportion taken by machinery from country i in I

y_i denotes the growth rate of labour productivity in country i on newly installed machinery

As is clear from the equation (2), the size of the positive impact of imported technology on the technological build-up of the recipient nation depends on:
a) its capacity to increase its stock of machinery
b) the installation lag (imitation lag)
c) the shares of different suppliers in the supplies of machines installed (vintage effect)
d) the ratio of each of these nations' incremental labor productivity to one of the recipient nations.

The concept sketched above apparently derives its basic logic from the capital vintage family of growth models, making use of the same mechanism as identified in the said models.

Much the same arguments were repeated subsequently by H.W. Singer and L. Reynolds, who also attempted to test the model statistically on a broader scale than Gomulka originally did. They concluded their exercise by saying ... "within the group of western style industrial countries the process of catching up by those with initially lower income per capita (their measure of technological gap - remark of the author) GNP explains a good deal of the differential increase in manufacturing sector. We associate this catching up process with the diffusion of technology" (7). Thus, the novum which they provided is the qualification restricting the application of the theory to "western style industrial countries".

A new dimension to this line of thinking has been added by R. Findlay, whose model of interaction between foreign technology and national technological development is based on two principal hypotheses:

1. the rate of technological change in a relatively backward region is an increasing function of the gap between its own level of technology and that of an advanced region (Veblen-Gerschenkron effect),

2. other things being equal, the rate of change of technical efficiency in the backward region is an increasing function of the relativ extent to which the activities of foreign firms, with their superior technology, penetrate the local economy (contagion hypothesis) (8).

Combining these two hypotheses, he formalizes his model as follows:

$$\dot{B}/B = f(x,y) \tag{3}$$

with $\partial f/\partial x < 0$ and $\partial f/\partial y > 0$

of which:

B is the technological level of the backward nation

$$x = \frac{B(t)}{A(t)} \quad \text{(relative backwardness)}$$

A is the technological level of the supplier

$$y = \frac{K_f(t)}{K_d(t)} \quad \text{(index of foreign investment penetration)}$$

K_f capital stock of foreign-owned (and managed) firms

K_d domestic-owned capital stock

In analyzing the Veblen-Gerschenkron effect, he postulates that

$$A(t) = A_0 e^{nt} \tag{4}$$

so that the technological level in the advanced economy increases at a constant rate n. If B(t) represents the relevant level in the backward country, then the Veblen-Gerschenkron effect can be presented as

$$dB/dt = \lambda \left[A_0 e^{nt} - B(t) \right] \tag{5}$$

where λ is a positive constant, its value depending on exogenous parameters such as the quality of management and the education of the labor force. Integrating (5) we arrive at

$$B(t) = \frac{\lambda}{(n+\lambda)} A_0 e^{nt} + \frac{(n+\lambda) B_0 - \lambda A_0}{(n+\lambda)} e^{-\lambda t} \tag{6}$$

of which B_0 is the initial level of technology in the backward region. With the time leading to infinity, (6) shows that the ratio of $B(t)$ to $A(t)$ will approach an "equilibrium gap" of $(n+\lambda)$ which varies directly with λ and inversely with n (9).

As far as the second hypothesis is concerned, R. Findlay makes use of the contagious disease analogy, the essential proposition of which is that technical innovations are most effectively copied when there is a personal interaction between suppliers and recipients. In the past, this has been acheived by a straightforward migration of foreign individuals. Today, however, their role - according to R. Findlay - has been taken over by transnational corporations (10). Hence, their importance to the technological catching up.

Thus, the new elements provided by R. Findlay are those which

1. relate to the "equilibrium gap", which indicates that the effects of technology import tend to phase out with the catching up process and therefore crossing the equilibrium gap requires different measures, and
2. his assumption concerning the positive function of foreign investment in the catching up process.

The question of the equilibrium gap was subsequently raised by S. Gomulka in his article published in 1985 (11).

In discussing this issue, he assumes that the innovation rate for any particular country tends to change with the level of technological development and it is the highest at the medium levels of development, i.e. when there is still a great deal to learn from other countries, and at the same time, proper infrastructural basis has been provided (education, R+D sector, investment goods sector, etc.). See Fig. 2.1.

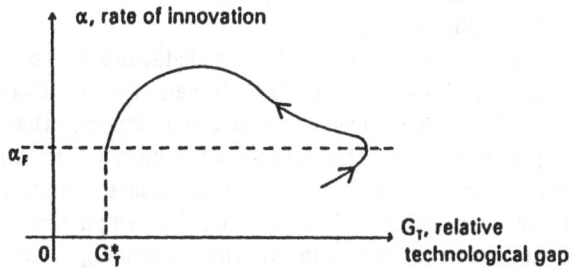

Fig. 2.1. The Growth Path of a Latecomer in the Course of Technological Catching-up, in Terms of the Innovation Rate and the Relative Technological Gap

Note: α_F denotes the rate of innovation assumed constant over time in the technological most advanced area, such as the US, for most of the twentieth century. The relative technological gap increases whenever $\alpha < \alpha_F$ and declines otherwise.

Source: S. Gomulka - the Incompatibility ... op.cit., p. 21

Gomulka argues that the magnitude by which the actual rate of innovation of the catching up country exceeds the equilibrium rate (the distance between α and α_F in Fig. 2.1.) is country specific and depends on the volume of resources allocated to capital accumulation and technological change and on the efficiency of their utilization. This, in turn, depends above all on a country's systemic and cultural characteristics. Countries with inefficient economic systems may still enjoy a high innovation rate, provided that G_T is greater than G_T^*, however, they must compensate for inefficiency with larger resources (12). With G_T approaching G_T^*, the innovation rate could be low and largely unresponsive to any further increase in the import of technology. Also, keeping G_T

close to G_T^* would require a continual large transfer of technology.

Thus, what is argued here, is that technology importation provides a stimulus to technological catching up only until G_T^* is approached. What is lacking in this argumentation, is what actually determines the distance between G_T and G_T^*, or, in other words, what determines the equilibrium gap.

A somewhat different perception of the relationship between catching up and the technology gap is offered by J. Cornwall (13). Assuming, like the other authors discussed above, that the crucial cause of rapid growth is the ability of countries to borrow technology from industrial leaders, which he associates with lower costs of imitation as compared to innovation, he emphasizes that this ability is not so much a function of the technology gap "but also and primarily of the existence of an entrepreneurial class keen and eager to exploit the available world stock of technology" (14). Thus, he draws attention to social and political factors as explanatory variables.

In their interesting paper published in 1981, K. Pavitt and L.G. Soete made an attempt to statistically test different variables for explaining the differences in output and productivity growth among OECD member states for the period of 1890-1977 (15). They found that the "catching up" variable (measured in terms of each country's productivity level compared to that in the US) had a positive impact on productivity growth from 1890-1913 and 1950-1970. In 1971-1977, however, the evidence is no longer clear and in some instances suggests a tendency towards divergence (16). In explaining this curious behavior of the catching up variable, the authors basically assign it to the increased homogenity of the technological level of the OECD countries (i.e. phasing out of the technology leads and lags) and thus the deteriorated prospects for the innovation profit appropriation by innovating organizations.

None of the theories briefly sketched above paid any special attention to the mechanism through which imported technology influences domestic technological performance. Nor have they elaborated on transactional characteristics of both home and foreign developed technologies. In fact, they were implicitly assuming that the two are basically the same phenomena. This is undoubtedly their severe drawback.

3. Resonance Model

This assumption has been challenged by the young Polish scholar W. Brzost, and subsequently by the author himself (17).

In investigating the transactional characteristics of imported technology, W. Brzost pays special attention to the differences existing between endogenous and exogenous (imported) technology. He basically views these differences as a result of the fact that the latter has been created outside the economic system in which it is subsequently being applied. Thus, the applying agent has a much broader scope of choice of the required technology, both as far as strictly engineering parameters and economic parameters are concerned. Thanks to this, the actual technological menu of a country may be considerably diversified and it is limited only by the current international supply of the technologies in demand.

The fact that one deals with imported technology also allows the recipient agent a high degree of freedom, as far as the level of the applied technology is concerned. Brzost argues that national technology is to a large extent strongly associated with the over-all national technological base, which is difficult to change in short time period, whereas in technology import there is a chance for jumping over certain development levels and cutting short development trajectories. This may be particularly rewarding in the case of newly established branches and product lines, while in traditional areas this is conditioned by the necessity to integrate imported technology into the existing technological framework at home.

Imported technology saves national R+D resources and, in most cases, also overall outlays which may be attributed to the price levels existing on the technology market. It also decreases the technological as well as the marketing risk, as, in most cases, the technologies become tradeable only after their successful implementation in the supplying country (18).

When borrowing technology abroad, the recipient nation can usually freely determine the intensity of the new technology inflow as well as the size and direction of the said inflow (given, of course, the lack of possible external limitations such as the availability of technology, availability of foreign exchange, etc.) which is largely restricted when only local sources of technology are opened.

A particularly interesting aspect of Brzost's analysis relates to the impact of imported technology on national technological progress. Remaining basically within the technology gap framework, he elaborates a model of economic resonance to explain the relationships between foreign technology and domestic technological growth (19).

According to Brzost, the effects of technology importation on national technological progress depend foremostly on three critical variables:

a) the intensity of the technology import and its structural composition,
b) the intensity of proinnovative forces operating in the recipient country,
c) the intensity of innovative obstacles present in the recipient country (20).

Intensity of
Technology
Import

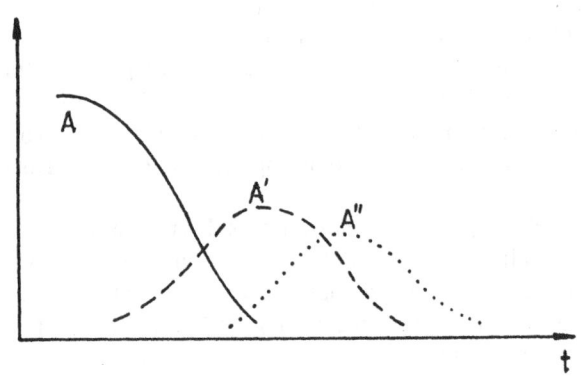

Local Technological Level

A - Effects of initial import of technology, t - Time
A', A" - Induced effect

Fig. 2.2. Technology Import and Resonance Effect, according to W. Brzost (21)

48

The intensity of the technology import is characterized by two variables: its superiority with respect to the local technological base, and the volume of the import. In principle, the higher the level of sophistication of the imported technology and the higher the average level of the local technological base, the larger the positive impact of importation on national technological progress.

Foreign technology introduced into the national economy represents a certain potential "technological energy" which is unloaded during the course of the implementation process. It is reflected in the increase of the technological level, up to the point which is represented by foreign technology. At the same time, however, local innovative acitivity is infected providing subsequent "technological energy" to be unloaded during the course of implementation processes.

This infection of local innovative activity results largely from both the new technological demands and opportunities brought forward by foreign technology. Normally, each new stimulus, being a sort of echo of the former one, represents a smaller unit of "technological energy", and thus the induced effects disappear with time. As such, the entire process of inducement - adaptation - inducement ... very much resembles the resonance effect well known in the area of mechanics (see Fig. 2.2.).

It is possible that in some cases induced technical change may outpace the original "technological energy" provided by the imported inputs. All in all, however, the existence of the said resonance effect results in higher effects of technology borrowing than is originally indicated by its face value (as presented in Fig. 2.3.).

There are however, as Brzost points out, two limitations which may distort the entire process. Both of them are related to the parameters of the importing country. The first one is its "reaction sensitivity" which determines the minimum import required to set the resonance process in motion. If this minimum import requirement is not met, then the imported technological energy is wasted. On the other hand, one is also confronted with the country's adapting capacity which determines the maximum ability of the importer to handle foreign technology. If this adapting capacity is counteracted (for example, by the import of too sophisticated technology or by a too broad stream of foreign technology or by too many changes in the composition of the imports) the negative resonance effect may appear (22).

Local
Technological
Level

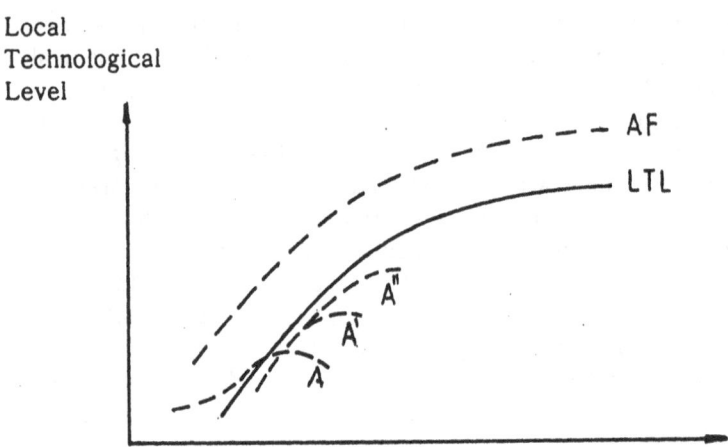

A, A', A" as in Fig. 2.
LTL - Local technological level
AF - Adaption frontier

Fig. 2.3.   Technology Import and Local Technological Level,
according to W. Brzost (23)

Thus, it is argued that the importer's maneuverability is limited
by these two variables. On the other hand, the variables are
country-specific and determined by the intensity of both the
forces stimulating and hampering technology in the given economy.
Increasing stimuli for innovation and/or for removing hampering
forces may provide the possibilities for both an increase in the
intensity of the import as well as an increase in the benefits received (see Fig. 2.4. where b_{max} and i_{max} represent respectively
increased resonance effects and increased adapting capacity). Brzost argues that the stimulating and hampering forces are foremostly the result of the developmental level achieved so far, the
systemic arrangements prevailing in the borrowing country, the industrial strucure of the economy, the size and composition of the
R+D sector, as well as the scope of the country's involvement in
the international division of labor (25).

50

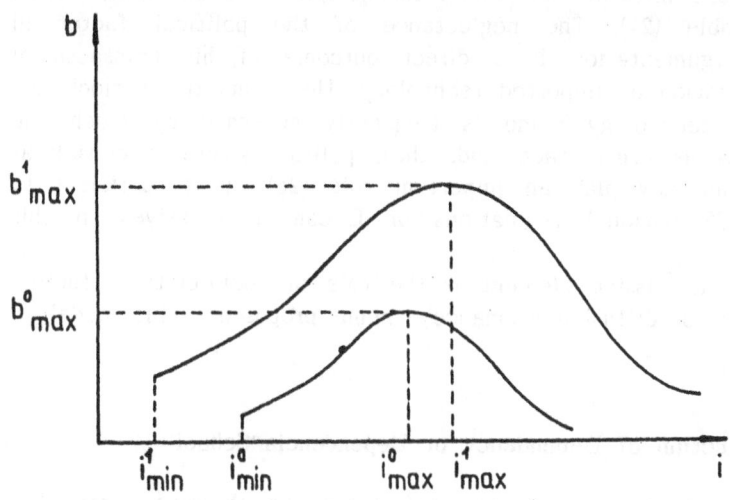

i - Intensity of technology import
i_{min} - Minimum import required to produce resonance effects
b - Technological benefits accredited to foreign technology

Fig. 2.4. Intensity of Technology Import and Resonance Effects, according to W. Brzost (24)

The model proposed by Brzost seems to considerably refine the original catching up/technology gap idea, by drawing attention to the various dimensions of interaction between foreign and domestic technologies, which have been largely neglected by his predecessors. It fails, however, to recognize some important factors deeply involved in the technology transfer process. The first one is the role of transfer channels applied. One can argue that the "technological energy" introduced by foreign technology will vary depending on the way it is being imported. Compact technological systems (for example turn-key installations) or the compact nature of importers (for example TNCs) may significantly alter the financial effects of the import operation (26).

The second one is the role of the international political setting or, more generally, the international environment which may have an important influence on which technologies and on which terms are available (27). The neglectance of the political factor in Brzost's argumentation is a direct outcome of his transactional characterization of imported technology. He seems to overlook the fact that technology trade is frequently governed by much the same rules as arms trade and, thus, political/strategic considerations undoubtedly play an important role (28). It is sufficient to mention US national regulations or Cocom prerogatives in this area.

The third missing element is the role of socio-cultural factors in the infusion-diffusion-resonance scheme proposed in his model.

4. Reproduction of Dependency or Dependencia School

In contrast to the technology gap/catching up theories, which basically viewed technological underdevelopment as a kind of economic resource and thus blessed the actions aimed at the utilization of foreign developed technology, dependencia thinkers were much more careful in this respect and frequently argued for quite opposite actions.

Unlike the technology gap models which were elaborated in the developed world, the dependencia school originated in the developing countries and clearly restricted its conclusions and recommendations to this group of countries. Thus, it did not claim universal applicability (as the former implicitly did) but insisted on its validity in the specific context of developing countries. Originally, it emerged from the disappointment among the developing countries with the results of industrialization processes carried out in the fifties and the sixties, and contributed substantially to the discussions on the new international economic order so widely spread in the mid and late 1970s. Its leading proponents came predominantly from Latin America, including such known personalities as A. Herrera, C. Furtado, F. Sagasti, R. Prebisch and, hence, its subsequent labelling as the dependencia theory. Later, however, it was further developed by radical economists and other social scientists all over the world (29).

The principal contention of this theory - although some believe is incorrect for this label (30) - is that the import of foreign technology from the developed countries does not serve the purpose of decreasing technological dependence but results rather in perpetuating and strengthening it. The most radical version of this theory argues that Western developed technology leads to the suppression of the local ability to produce technologies, thus perpetuating the dependency syndrome (31).

The theoretical foundations of the dependency theory are based on the infant industry arguments advocated long ago by a German economist, F. List, and the theory of the dominant economy of the French economist, F. Perroux. F. List argued that the free trade presupposes an earlier creation of a healthy industry at home, which can be achieved only by the erection of protectionist barriers during the infant phase (32). F. Perroux, on the other hand, points out that the world economy is composed of dominating and dominated economies, resulting in specific types of relationships between the two groups of nations (33). This, in turn, may result in the appearance of the domination effect, including inter alia, some forms of economic exploitation of the dominated economies.

Dependency theorists argued that both the nature of technologies which are available for the developing countries, as well as the contractual conditions prevailing, are highly inconsistent with the needs and possibilities of the developing world.

As far as the first issue is concerned, it is claimed that the technology available in the developed world is generally inappropriate, strictly from a production point of view. It reflects the actual resource endowment of the developed world and not that of the developing nations. Thus, as a rule, it is too capital-intensive and too little labor-intensive. In the end, it reinforces the structural unemployment characteristic for the underdeveloped world and overconsumes the limited capital resources (34).

It is, however, also inappropriate from a more general social point of view, as it brings along unwelcome elements of Western culture, diverting economic resources from agricultural production and overemphasising the import of luxury products (35). Even if appropriate technologies are available, multinational companies are rather reluctant to provide them, not to mention the adaptation necessary unless they are forced to do so (36).

It is, however, not only the nature of the offered technologies which matters, but also the commercial conditions to which the technology becomes available. Here, dependency thinkers point to the facts (along the line of Perroux's theory) that developing countries are subject to numerous contractual limitations, as far as the utilization and further development of imported technology is concerned, they are subject to unduly high prices, forced to acquire packaged technologies, and so forth (37). All of these drain their scarce economic resources and hamper local technological learning.

Thus, the most radical dependency theorists argue that, "Instead of looking towards the transfer and the adoption of developed countries' technologies as the solution of their problems, third World countries should adopt an autocentered development strategy combined with the development of autochthonous technologies, tailored to their conditions" (38). On the whole, however, the policy recommendations are not that extreme, and call rather for a development rationale based on the selective dissociation with developed countries and strong governmental involvement, both in the creating of a local technological base and in the process of acquisition of foreign technology (selection, improvement of bargaining position, diffusion activities, etc.) (39).

A major criticism of the dependency theory is that it is basically negative in its argumentation and does not provide any reasonable alternatives. So, while having said what not to do, it has failed to suggest positive solutions to the problems (40).

This seems to be a severe underestimation of the said theory which, by all means, has played an important role in broadening and deepening our understanding of the relationship between the import of technology and local technological development.

It challenged the universal validity of the technology gap/ catching up approach, and drew special attention to international limitations of technology flows. It stressed the asymmetric power relations in the contemporary world and their influence on the international flow of technology. It drew due attention to local technological capacity as a precondition for a smooth intake of foreign developed technologies. Last but not least, it pointed out the necessity of local technological production in all those areas which the international community seems to neglect, in view of different social, political and economic priorities.

It should be stressed that the dependency theory heavily influenced policy makers in the developing countries, being at the root of national regulations on technology import initiated in the 1970s, of negotiations on the International Code of Conduct for Technology Transfer, as well as of attempts for the establishment of South-South technological cooperation.

5. Foreign Technology and National Survival: The Case of the Socialist Countries

A separate line of theoretical debate on the relationship between technology import and national technological development is concerned with the issue of the socialist countries of Eastern Europe. It is interesting to note that the views and arguments offered in this debate seem to basically overlook the arguments developed by other theoretical models and live their own lives, concentrating exclusively on the institutional features of the centrally planned economies as the major variable of their considerations.

The standard observation of the said theorists, which constitutes their point of departure, is the apparent inefficiency of the centrally planned economies to create, implement and diffuse technological change. Therefore - it is argued - there is a need for technology import from the industrialized West to compensate for this inefficiency. Some of the authors view this importation as a means of accelerating the technological catch up and not necessarily as the condition of sine qua non of the technological development of the said countries (41). Some others, however, claim that foreign technology is a life preserver for otherwise clumsy economic systems of the centrally planned economies (42).

The two most well known proponents of this idea are the German scientists Jochen Röpke and Siegfried G. Schoppe.

In his paper published in 1976 under the title - "Imported progress: import of innovation as a survival strategy of centrally coordinated systems", J. Röpke develops the following line of argumentation (43):

a. The content of new knowledge is unpredictable. The results of the production of new knowledge are uncertain. The production of new knowledge is based on "real" decisions which cannot be based on the calculation of maximization or optimization functions.

b. If it is uncertain where and when new ideas emerge and what qualities they have, then, it is clear that the production of new ideas cannot be centrally coordinated, as the central coordination requires prior knowledge of the results of the undertaking.

c. Growing complexity of the system increases the range of "real" decisions and hence, the necessity for the decentralization of decision making and coordinating processes.

d. As a result, monocentric systems (i.e. centrally planned economies - J.M.) acquire new technological knowledge, either from abroad or from non-centrally coordinated poles of the national economy.

e. As non-centrally coordinated poles represent, as a rule, an unimportant part of the national economic systems, technology import from polycentric system (read - market based economies - J.M.) are necessary for monocentric systems to survive in the international community (42).

These rather simplistic arguments of J. Röpke become much more refined and elaborated in the works of S.G. Schoppe (45). The subject of relationships between imported and home grown technologies was not a central theme of Schoppe's writings. He was largely concerned with the explanation of the structural composition of the East-West trade, offering an interesting theory on the propensity to risk-determined advantages (46). In the course of his investigations, however, he too embarked on the issue of technology import to the centrally planned economies. Remaining largely within the framework of the new institutional economy, which attempts to explain economic activities by referring to the institutional characteristics of the given systems and the transactional characteristics of the specific activities, he pays particular attention to the pertinent coordination mechanism, prevailing in the hierarchical (centrally planned economies) systems.

He argues that hierarchical systems, plagued by a centralized coordination mechanism, are characterized by a high degree of risk avertion (or in other words, they demonstrate a low propensity towards risk bearing and a low capacity for risk distribution). Therefore, they tend to reject innovations as risky undertakings, which inject undesirable uncertainty into the hierarchy (47). This risk aversion, coupled with the lack of entrepeneur-property motivations (collective property is no one's property), inefficient centralized information systems (Röpke's hypothesis) and centralized ex-ante allocation of resources, finally results in the fact that the technology generation within hierarchical systems is largely ineficient. The same is true for the technology transfer among hierarchical systems (48). On the other hand, the same processes in market type economies develop in a largely unhampered way.

Logically, he then draws his final conclusion by saying: "In case of lack of any interactions between market type and planned type economies the latter will be continuously slipping back in their technological development. When technology transfer from market type economies is undertaken it is possible to limit the rise of the technology gap. Technology import, however, fails to bridge the relevant gap (permanent technological gap) for the smooth diffusion of external technologies due to systemic barriers. Only in case of (a) decentralization of the system (read - change of the system - J.M.) or (b) creation of the parallel economy (read - technology islands outside systemic conditions - J.M.) there is a chance for closing the technology gap in foreign trade" (49).

In other words, to survive internationally, centrally planned economies are condemned to the importation of Western technology. As we can see from the foregoing analysis, the said author is not particularly interested in the mechanism of interaction between imported and home grown technologies, but considers the former as the substitute for the lack of the latter.

The major contentions of this theory may be challenged on two grounds: empirical and theoretical and methodological. Let us take the first dimension. First, evidence abounds that market systems alone do not provide an automatic solution for the technological gaps, which in an overwhelming part of the Western world are - so far - of a permanent and not - as suggested by Schoppe - of a transitional nature. Hence, other factors seem to be equally important. Second, it remains to be proved whether the importation of Western technology provides significant impulses for technological as well as overall economic acceleration in the centrally planned countries. Particularly, it must be seen whether an increase in the volume of importation results in the relative decrease of the technology gap. The findings produced so far provide a rather negative answer to this assumption (50).

As far as the second dimension is concerned, let us change the formulation of Schoppe's hypothesis slightly to see the results. It would then read as follows: "In case of lack of any interactions between technologically dominant nations and those which lag behind the latter will be continously slipping back". In other words, the hypothesis which is offered seems to be equally valid for any type of economic system and, moreover, it can be equally valid for any country whatever its initial technological position. Planned economies generally do not become technologically obsolete when they became hierarchical systems. They were rather born as technological laggards. This argument, therefore, should not be misused.

These reservations are not to be taken as a defence for the apparent low efficiency of the socialist countries in technology creation, transfer and utilization. Nor are they meant as a total disqualification of some of Schoppe's propositions. They are rather meant to put the phenomenon in a proper perspective and point out the diversity of variables which should be taken into consideration.

6. Conclusions

As can be seen from the preceding discussion, the relationship between imported technology and national technological build-up is far from being identified and clarified. In particular, little is known about the nature and determinants of the linkage effects between foreign and domestic technologies, the role of international politics, the economic and technological set up, the transactional characteristics of various transfer channels, or the impact of socio-cultural and policy related variables. The theoretical models reviewed remain on a fairly high level of abstractness and do not - as a rule - embark on a more detailed debate of the issues in question. Thus far, theoretical concepts seem to deal with each of these separately, addressing only a part of the overall truth, which is much more complex and multidimensional.

At the same time, however, it seems that these concepts do contain some basic premises for the subsequent elaboration of the more refined approach. They point, first of all, to the importance of the initial level of the overall economic and technological development for the opportunities and limitations of the import-led technological growth. This is, as a matter of fact - paradoxically as it may be - a common denominator of both the production of development and reproduction of dependency approaches, which, departing from much the same observations, arrive at totally different conclusions. They indicate too - although to a much lesser extent - the importance of the national technological and economic policy of the importing nations. Increasing or decreasing the inflow of foreign technologies, allowing or forbidding the use of certain channels, shaping the structural composition of the imported technology, etc. will finally be articulated in the results recorded in the course of the catching up process. Finally, they draw our attention to the role of the systemic or institutional factors which are constituent of the way the given economic system operates and thus the way it absorbs, diffuses and links foreign technology with local innovating activities.

What they apparently fail to recognize is the paramount importance of the socio-cultural elements which, to a considerable extent, determine the receptivity of the society, its attitude towards foreign technology and its functioning within the given institutional framework (51).

They also seem to neglect the role of an international set up for the technology transfer process, both from an economic as well as a political view. As indicated before, technology trade displays much the same characteristics as the trade in arms. Thus, the current political and strategic atmosphere shapes the attitudes of both the donor and the recipient countries. On the other hand, technology as an element of economic power requires an application of the theory of economic power to explain the uneasy relationships between the suppliers and the receivers of technology. Only when these elements are taken together a proper framework for the explanation of the phenomenon in quest is provided.

References and Footnotes

1   Ch. Cooper - Science, technology and production in the underdeveloped countries: an introduction, The Journal of Development Studies, Vol. 9, October 1972, no. 1, pp. 1-2
2   This is particularly marked in the programs of the United Nations Conference on Trade and Development and the United Nations Industrial Development Organization
3   T. Veblen - Imperial Germany and the industrial revolution, London, 1915. A. Gerschenkron - Economic backwardness in historical perspective, Cambridge, 1962
4   A. Gerschenkron - Economic backwardness in historical perspective, Cambridge, 1962, p. 8
5   S. Gomulka - Inventive activity, diffusion and the stages of economic growth, Aarhus University Press, 1970
6   Taken from P. Hanson - Trade and technology in Soviet - Western Relations, The Macmillan Press Ltd., 1981, p. 23

7   H.W. Singer, L. Reynolds - Technological backwardness and productivity growth, The Economic Journal, March 1975, p. 876

8   R. Findlay - Relative backwardness, direct foreign investment and the transfer of technology: a simple dynamic model, the Quarterly Journal of Economics, February 1978, no. 1, pp. 1-16

9   R. Findlay - Relative backwardness ... op.cit., p. 3

10 R. Findlay - Relative backwardness ... op.cit., p. 4

11 S. Gomulka - The incompatibility of socialism and rapid innovation, in: M.E. Schaffer (ed.) - Technology transfer and East-West relations, Croom Helm, 1985, pp. 12-31

12 S. Gomulka - The incompatibility ... op.cit., p. 22

13 J. Cornwall - Modern capitalism. Its growth and transformation, London, 1977

14 Quoted from K. Pavitt, L.G. Soete - International dynamics of innovation, in H. Giersch (ed.) - Emerging technologies: Consequences for economic growth, structural change and employment, J.C.B. Mohr, Tübingen, 1982, p. 122

15 K. Pavitt, L.G. Soete - International dynamics ... op.cit., pp. 105-133

16 K. Pavitt, L.G. Soete - International dynamics ... op.cit., p. 126

17 W. Brzost - Importowany postęp techniczny a rozwój gospodarczy Polski, PWN, Warszawa 1979; J. Monkiewicz - Międzynarodowy transfer wiedzy technicznej; elementy teorii i polityki, PWN, Warszawa 1981

18 W. Brzost - Importowany postęp techniczny ... op.cit., p. 182

19 W. Brzost - Importowany postęp techniczny ... op.cit., pp. 179-197

20 W. Brzost - Importowany postęp techniczny ... op.cit., p. 190

21 W. Brzost - Importowany postęp techniczny ... op.cit., p. 187

22 W. Brzost - Importowany postęp techniczny ... op.cit., p. 188 See also J. Monkiewicz - Międzynarodowy transfer wiedzy technicznej ... op.cit., pp. 82-83

23 W. Brzost - Importowany postęp techniczny ... op.cit., p. 188.

24 W. Brzost - Importowany postęp techniczny ... op.cit., p. 193

25 W. Brzost - Importowany postęp techniczny ... op.cit., pp. 191-192

26 J. Monkiewicz - Międzynarodowy transfer wiedzy technicznej ... op.cit., pp. 31-63

27 L.A. Nefiodow - The effects of East-West conflict on international competition for high technology, mimeo, undated

28 J.S. Fitch - International transfer of military technology: the political impact of US military aid to Latin America, Dept. of Political Science, University of Colorado, mimeo, undated

29 D. Ernst (ed.) - The new international division of labour, technology and underdevelopment. Consequences for the Third World, Campus Verlag, Frankfurt/New York, 1980

30 A. Emmanuel - Appropriate or underdeveloped technology? John Wiley and Sons, 1982, p. 30

31 L. Mytelka - Regional development in a global economy: the multinational corporation, technology and Andean integration, New Haven, Yale University Press, 1979

32 F. List - Der internationale Handel, die Handelspolitik und der deutsche Zollverein, in: F. List - Das nationale System der politischen Ökonomie, Jena 1920

33 F. Perroux - L'êconomie du XXe siêcle, Presses Universitaires de France, 1961; F. Perroux - The domination effect and modern economic theory, in: K.W. Rotschild (ed.) - Power in economics, Penguin, 1971

34 K. Poznanski - Technology transfer: West-South perspective, World Politics, Vol. 37, no. 1, 1984, p. 145

35 J. Galtung - Towards a new international technological order, Alternatives, 4 (1978-79), pp. 277-300

36 C. Vaitsos - Government policies for bargaining with transnational enterprises in the acquisition of technology, in J. Ramesh, Ch. Weiss (eds.) - Mobilizing technology for world development, N.Y., Praeger, 1979

37 L. Mytelka - Regional development ... op.cit., pp. 129-135

38 T. Baumgartner - Transfer of technology: production of development or reproduction of dependency, in: D. Ernst (ed.) - The new international division of labour, technology and underdevelopment, Campus Verlag, Frankfurt am Main/New York, 1980, p. 581

39 S. Rosenblatt (ed.) - Technology and economic development: a realistic perspective, Westview Press, 1979

40 A very thorough criticism may be found in A. Emmanuel - Appropriate or underdeveloped technology. J. Wiley and Sons, 1982

41 A leading proponent of this argumentation is P. Hanson, see: P. Hanson - Trade and technology in Soviet-Western relations, The Macmillan Press, 1981

42 P.V. Elst - Capitalist technology for Soviet survival, Institute of Economic Affairs, Great Britain, 1981

43 J. Röpke - Der importierte Fortschritt. Neuerungsimport als Überlebensstrategie zentralkoordinierter Systeme, ORDO, Jahrbuch für die Ordnung von Wirtschaft und Gesellschaft, Vol. 27, pp. 223-241

44 J. Röpke - Der importierte Fortschritt ... op.cit., p. 235

45 S.G. Schoppe - Die sowjetische Westhandelsstruktur - ein außenhandelstheoretisches Paradoxon, Gustav Fischer Verlag, Stuttgart, New York, 1981; S.G. Schoppe - Koordinationsmechanismen beim nationalen und internationalen Technologietransfer (TT) der Sowjetunion anhand konkreter Beispiele, offprint, undated; S.G. Schoppe - Die intrasystemaren und die intersystemaren Technologietransfers der DDR, in: Schriften zum Vergleich von Wirtschaftsordnungen, Heft 30, Gustav Fischer Verlag, 1983, pp. 345-362

46 S.G. Schoppe - Die sowjetische Westhandelsstruktur ... op.cit.

47 S.G. Schoppe - Die sowjetische Westhandelsstruktur ... op.cit., pp. 80-81

48 S.G. Schoppe - Die sowjetische Westhandelsstruktur ... op.cit., p. 79

49 S.G. Schoppe - Die sowjetische Westhandelsstruktur ... op.cit., p. 79

50 See particularly the series of reports published recently by OECD: E. Zaleski, H. Wienert - Technology transfer between East and West, OECD, 1980; Z. Fallenbuchl - East-West technology transfer. Study of Poland 1971-1980, OECD, 1983; S. Gomulka, A. Nove, G.D. Holliday - East-West technology transfer, OECD, 1984; F. Levcik, J. Skolka - East-West technology transfer. Study of Czechoslovakia, OECD, 1984; M. Bornstein - The transfer of Western technology to the USSR, OECD, 1985

51 E.M. Rogers - Diffusion of innovations, New York, 1962; E.M. Rogers, F.F. Shoemaker - Communication of innovations. A cross cultural approach, The Free Press, 1971

Chapter 3
The Theory of the Dominant Economy and International Technology Transfer

1. Introduction

As pointed out in the preceding chapter, one of the important variables explaining the technology transfer process is the international political and economic set up, within which the entire phenomenon takes place. This seems to be particularly relevant with respect to East-West relations. Interestingly enough, however, this issue has not been in the forefront of the analyses in the existing literature, at least in a more disciplined manner. The aim of the present chapter, therefore, is to fill this gap by applying the analytical framework of the theory of economic power and of the dominant economy, developed by the French economist Francois Perroux.

The issue of power has not been a favorite subject of economic writings so far. As properly observed by K.W. Rothschild: "Yet, if we look at the main run of economic theory of the past hundred years we find that it is characterized by a strange lack of power considerations. More or less homogenous units - firms and households (and one could add countries - J. Monkiewicz) - move in more or less given technological and market conditions and try to improve their economic lot within the constraints of these conditions. But that people will use power to alter the mechanism itself; that uneven power may greatly influence the outcome of market operations, that people may strive for economic power as much as for economic wealth: these facts have been largely neglected" (1).

It is not intended to analyze here the reasons for such a situation but rather to see what could be achieved in the understanding of international and especially East-West technology transfer if the element of power is inserted into the analysis.

In the following, I shall briefly recall the major elements of Perroux's power-centered theory of the dominant economy and thereafter I shall attempt to apply it to the analysis of the East-West technology relationships.

2. The Theory of the Dominant Economy: The Constituents

The theory of the dominant economy has been elaborated in the late 1940s and early 1950s by Francois Perroux as a reflection on the US domination of the Western world (2). Drawing on the earlier works on economic power of E. Böhm-Bawerk and T. Veblen on the one hand, and on imperfect competition (in particular E.H. Chamberlain) on the other, he produced his own original theory which aimed at explaining the causes and effects of unequal distribution of power among economic agents.

The principal thrust of the theory is perhaps best characterized with the following quotation from one of his essays: "Economic life is something different from a network of exchange. It is rather a network of forces", and further: "The economy is guided not only by the search for gain but also by that for power" (3).

The point of departure of this theory is an assumption that the state of equilibrium of exchange, according to L. Walras and V. Pareto, represents merely a special case of a much larger cluster of disequilibrium of different types which tend to exist constantly and not just transitionally (4). This is particularly true with respect to international trade, which means that unequal exchange may be something quite normal and not extraordinary.

The principal elements of Perroux's theorem are the dominant economy and the domination effect. To explain them let us begin with the model of economic relations between two units (nations) - A and B (5).

a) Let A influence B with the force n_1

$$A \longrightarrow B = n_1$$

b) Let B return this initial influence of A with n_2

$$B \longrightarrow A = n_2$$

c) Now if

$$n_1 - n_2 = 0$$ i.e. the two units (nations) are in equilibrium (there is no dominant economy and no domination effect emerges)

$$n_1 - n_2 = +n_1 = \gamma$$ i.e. A is dominating B and γ is the net domination effect

$$n_1 - n_2 = -n_1 = \sigma$$ i.e. B is dominating A and σ is the net domination effect

Thus, we see that the term "dominant economy" is used to describe unequal power relationships between (among) units (nations) under consideration. The very existence of the dominant economy may be considered as biased interdependence and the <u>dominant economy could be defined as the economic system exerting stronger impacts on the individual economies concerned than their actual counter-impacts.</u>

It means that the actual dependency of the said nations cannot be described with the following formula (6):

$$N = f(Z); \qquad Z = f^{-1}(N)$$

It should perhaps be noted, that the term "dominant economy" is a relative one and that actual economic activities take place among the economies which could be simultaneously both dominant and dominated, forming a complex system of domination and dependency world-wide.

There can be a universally dominant economy (for example U.K. in the 18th and 19th centuries, the USA in the postwar period), regionally dominant economies (for example Brazil in Latin America), or locally dominant economies (for example the FRG and Austria).

<u>The domination effect is a result of the asymmetrical influence of one economy (or group of them) on the other</u> (7). It could be measured with the degree of deviation from the state of equilibrium of exchange, which is characterized by the following:

$$\frac{\text{Marginal advantages received (MAR)}}{\text{Marginal advantages handed (MAH)}} = 1$$

If this relationship for a given economy $\neq$ 1, it means that this economy is either dominant (MAR / MAH > 1), or dependent (MAR / MAH < 1) (8).

The domination effect may be either non-intentional (liaisons non-intentionelles) or intentional (liaisons intentionelles), depending whether it is the result of purposeful actions aimed at its accomplishment, or just the side effect of autonomous actions taken by the given economy and justified by quite different considerations. For instance, if due to its structural policy a given economy, A, changes its economic structure and thus brings about the structural adaptations in B, then the non-intentional domination effect occurs. If the same action is taken in A, with the aim of restructuring the economy of B, we are confronted with the intentional domination effect.

According to F. Perroux, there are three sources (or constituents) of the domination effect, apart from the strictly military presence (occupation or colonization). These are:

a) differences in bargaining power among the nations (force contractuelle, pouvoir de négociation),
b) differences in economic size of the actors (dimension), and
c) differences in the nature of activities performed (nature de l'activité) (9).

One should note perhaps, that the differences in the bargaining power are foremostly based on the size and nature of the economic activities (especially the type of key resources possessed) of the given economy, as well as on its military position. The bargaining power is manifested in the ability of setting up or transforming the rules of the game (transformation des règles du jeu).

The domination effect, according to F. Perroux, is neither overriding nor unlimited and it would cease with the fall of the dominant economy. It would be revitalized, however, with the emergence of new dominant centers.

The process of the development of the world economy is - as pointed out by Perroux - determined not by the competition among the equal partners but by the subsequent rise and fall of major national economies of a dominant nature (10).

The dominant economy may exert either a pull effect (effet d'entrainement), or a stoppage effect (effet de stoppage), by providing or rejecting the dominated economies access to its market and other key resources (technology, equipment, critical raw materials, education system, etc.). Thus, it is very important what type of relationship prevails between the dominant center and its individual dominated partners.

Let us now depart from the previous assumption of a two-nations world and introduce the element of alliances (interest groups, i.e. economic groupings). In the case of a multi-nation world the appearance of the domination effect becomes more complex and, in addition to the direct effect discussed so far, we might also be confronted with indirect effects.

Assuming that we are operating in a world composed of three countries, we could have the following situation:

1. A influences B;   A $\longrightarrow$ B

2. B counter-influences A;   B $\longrightarrow$ A

3. B influences C;   B' $\longrightarrow$ C

4. C counter-influences B;   C $\longrightarrow$ B

The net effects of 1. and 2. could then be called the direct, and of 3. and 4. the indirect domination effect of A. Of course this type of chain-analysis could continue almost indefinitely.

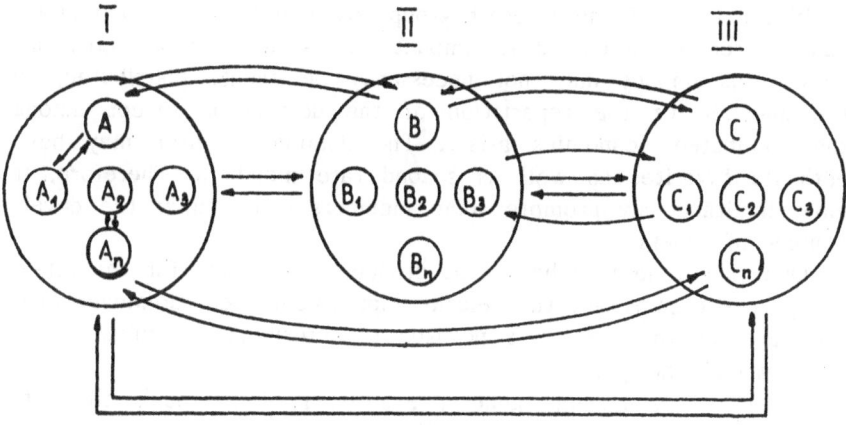

Fig. 3.1.   The network of domination/dependency relationship in multi-nation world with various interest groups

The issue becomes even more complicated if we assume that A, B and C are all members of some interest group I, II or III, and that their individual relationships are governed both by their individual inclinations and properties as well as by their group membership.

The group membership sets forth some sort of externalities (Rahmenbedingungen) for individual member-states. F. Perroux associates this particularly with the different impact power of individual groups vis-à-vis the external world and with the group-specific information supply (12). It seems, however, that equally important are the existence of some internal rules of behavior (degree of discipline exercised by the group and particularly by the dominant center) and the presence of common interests among the group members.

There might be significant differences among different groups, as far as internal discipline and thus homogenization of members' actions and reactions are concerned. This relates principally to the position occupied by the group dominant center. It may further evolve over time, providing more or less freedom for the individual member states in their external relations.

Within each of the interest groups we can have a group dominant center which exerts its impact on its allies (group co-members). This impact may not necessarily be evenly distributed, so the question of the repartition of the domination effect among the dominated economies arises. The dominant center may have both its favorites as well as second rate members. Therefore, it may simultaneously promote some members and hamper the development of others.

One should note perhaps that critically important for the inter-group relationships are the relationships among local dominant centers, as only they are in a position to determine the rules of the game within the group.

After this somewhat general exposition of Perroux's theorem, let us now look closer into East-West technological relationships.

3. International Setting for East-West Technology Transfer

East-West technology transfer takes place under specific conditions characterized by the following major features:

a. incompatibility of political and ideological aims and values of the actors concerned,
b. unequal power relationship of the parties, both with respect to the size as well as to the nature of their economic activities,
c. presence of organized political-economic groupings on both sides, with clearly outlined local dominant centers,
d. military rivalry between local dominant centers: US on the one hand and the Soviet Union on the other.

The incompatibility of the political and ideological aims and values of the parties in question is not a matter of secondary importance but clearly of fundamental nature. Political and ideological aims and values as visualized by the socialist countries of Eastern Europe (i.e. the East) are considered by industrialized capitalist countries (i.e. the West) as principally alien and subversive; they should not be simply taken into account but should be counter-acted. In spite of the peaceful co-existence rhetoric, the socialist countries are seen as the principal challengers of the world establishment (i.e. the West) and hence, should be - in their own interest - defeated on all grounds. Thus, we can say that the unavoidable and constant political and ideological animosity shapes the general framework for East-West relations.

The incompatibility of political and ideological aims and values has its consequences in the economic activities of the parties, including technology transfer.

Parallel to this, East-West technology transfer occurs within the framework of the unequal distribution of economic power between the parties concerned, of which some elements are shown in Table 3.1.

As we can note, both with respect to the size of the economic activities as well as to their nature, the Western countries have a significant edge over the Eastern countries. In terms of R+D expenditures, the amount spent in OECD countries is nearly 3 times higher than in CMEA, in terms of patenting abroad, the ratio is 27:1, in terms of the share in world trade, the ratio is about 7:1, in terms of the mutual trade dependency, the ratio is about 7,5:1 (i.e. the Eastern share in OECD trade is over 7 times lower than the OECD share in CMEA trade flows).

Economic relationships among individual Western and Eastern countries are under the heavy impact of the existing organized political-economic groupings, NATO, OECD and EEC being the

Table 3.1. Elements of the Power Relationship between East and West

Indicators	CMEA (1)	OECD (2)	Coefficient of OECD/CMEA power relationship 2:1 (%)
1. Territory (1984) % of the world total	18.7	23.9	1.26
2. Population (1984) % of the world total	9.6	16.6	1.72
3. GDP per capita (1982) in $ 1000	2.6	9.6	3.69
4. Export per capita (1984) in %	468	1552	3.32
5. R and D expenditures (1980) % of the world total	24.4	72.7	2.98
6. Patents granted to residents abroad (1980) % of the world total	3.0	80.0	26.66
7. Mutual share in total (1982)			
a) exports	3.6	29.5	8.19
b) imports	4.1	28.6	6.98
8. Share in world (1983) in %			
a) exports	9.8	64.0	6.53
b) imports	8.9	65.9	7.40

Source: Rocznik Statystyczny GUS, Warszawa 1985, pp. 521-523; J. Monkiewicz, J. Maciejewicz - Technology export from the socialist countries, Westview Press, Boulder and London, 1986, p. 19 and 26; Handbook of International Trade and Development Statistics, Supplement 1985, UNCTAD, New York 1985, pp. 26, 72, 80, 92, 100, 416

leading representatives of the Western interests, and the Warsaw Pact and CMEA as their Eastern equivalents.

The NATO and Warsaw Pact respectively set forth the strategic and military-political boundaries for the mutual economic relations. OECD, EEC on the one hand, and CMEA on the other, organize and work out the strategic-economic lines of mutual economic relations for the countries concerned. All this takes place under the strong impact of the local dominant centers: the US in the West and the USSR in the East. Both use their ideological, military and economic position to provide the necessary element of discipline among the respective group members. At the same time, the two dominant centers assume final responsibility for the military security of the respective groups, thus carrying the main burden of the military rivalry.

4. The Theory of the Dominant Economy and East-West Technology Transfer

The preceding analysis has pointed out that East-West economic and technological relations take place in a position clearly dominated by the Western countries. The size and the nature of their activities provide a base for both the un-intentional as well as intentional domination effect. To begin with, one can say, that it is finally the Western developed countries which determine the rules of the game on the world markets and that the CMEA countries have no choice but to accept them, at least in their external relations. This is well illustrated by their partial adherence to such organizations as the General Agreement on Trade and Tariffs (GATT) or the International Monetary Fund. Once completely rejected, they were then considered apparently necessary for the existence of the CMEA countries within the international community (13).

This comes as no surprise if we consider the fact that CMEA countries are nearly eight times more dependent on OECD markets, in quantitative terms, than vice-versa. Such an asymmetrical dependence, even in purely quantitative terms, may provide enough leverage for both the intentional and unintentional domination effect, with respect to terms of trade, structural changes and the like.

Possessing an overall technological lead over the CMEA countries, the OECD member countries also undoubtedly exert an unintentional domination effect, with respect to the industrial structure and the direction of technological activities carried out within the CMEA. This unintentional domination effect would be present even in the case of complete isolation of the two groups, either through third countries (i.e. developing world) or/and through military actions. Competing on third markets requires respective adaptation measures and the same is true with the development of new weapon systems.

Possesssion of the overall technological lead may also be used to achieve the intentional domination effect, which, in fact, has frequently been the case.

Considering the antagonistic nature of the political and ideological aims and values of the East and West, we could logically assume that the final goal of the Western countries in their technology export policy vis-à-vis CMEA is not to assist the latter in their national and regional technological build-up but rather to achieve short-run economic advantages under the condition of preserving their long-run dominant position in technology.

This seems to have been safeguarded by a number of group-coordinated measures. The first and most publicized instrument is strategic export control, aimed at limiting the access of the CMEA countries to certain types of technology, considered as military important.

The contemporary history of such strategies goes back to the early 1920s, when the US Administration announced an economic blockade of Soviet Russia, followed thereafter by an internationally coordinated credit boycott. It was again the US government which initiated a strategic export control against the socialist countries in the late 1940s. To make it more effective, the US succeeded in setting up an informal Coordinating Committee for Multilateral Export Control (COCOM) in 1949. It currently embraces all NATO allies (except Iceland) and Japan. COCOM maintains agreed munitions, nuclear and industrial lists of items, the export of which to so-called communist countries is controlled. Certain exporting may be permitted by agreement within COCOM, however, it is still subject to monitoring and review (14). Officially aimed at prohibiting the transfer of military technology to the East, the embargo policy inhibits also - as indicated by the leading Western specialist in this area, P. Hanson - West-East technology transfer in general (15). The same author also points out that, in practice, the machinery of the strategic embargo has some relevance on the implementation of sanctions and economic warfare. This is - he says - "because the general state of East-West relations tend to influence:

(a) the proportion of tradeable goods covered by embargo lists
(b) the granting of exceptions relating to items on the lists, and
(c) the severity of monitoring of illicit flows" (16).

The overall impact of the strategic export control machinery on Eastern access to Western technologies is hard to assess in quantitative terms. On the whole, it does not seem to be very dramatic assuming that the final effect of strategic export control regulations would basically be reflected in the lower share of high-technology items in the Western exports to the East, compared to the world total (Table 3.2.).

As indicated in Table 3.2., the said difference for the USSR in 1982 was minus 3.4 percentage point for the total export and minus 3.7 percentage point for manufacturing export. For Eastern Europe (excluding the Soviet Union) the differences in 1982 were smaller and amounted to minus 1.3 percentage point for total export and minus 0.7 percentage point for manufacturing export. Hence, the relative high share of high technology items in Western export to the USSR was around 30% lower and, in the case of Eastern Europe, around 10% lower than the world average.

The recorded difference may be taken as an indication of the repartition of the domination, or more precisely, the stoppage effect exerted by the Western countries. It is clear that the principal target of the instrument applied is the Soviet Union, as the most important member of the CMEA system. Judging from the figures given above, the overall impact of this stoppage effect does not appear too dramatic.

However, this picture changes significantly if a more detailed analysis is undertaken. Thus, for example, if we look at the sources of Western high-technology supplies to the Soviet Union, we notice that most of these clearly come from secondary sources, non-COCOM members. The US position as a high technology supplier is, in fact, negligible (see Table 3.3.). The largest suppliers of high-technology items to the USSR throughout the 1970s were W. Germany, followed by Japan and Finland (!), Italy, France, Switzerland, the UK and Sweden. The United States provided only half of the amount delivered by Sweden and less than

Table 3.2. Comparison of High-Technology Exports with Manufactured Goods and Total Goods Exported - 17 I.W. Countries to the Communist Countries and to the World: 1970, 1980, 1981, 1982 (millions of US dollars)

I.W. Exports to:	1970 $	1970 % of	1980 $	1980 % of	1981 $	1981 % of	1982 $	1982 % of
USSR								
High-Technology	402.9		2330.3		1774.4		2145.7	
Manuf. Goods	2212.4	18.2	151132.1	15.4	14183.1	12.5	15641.3	13.7
Total	2490.8	16.2	19837.5	11.7	20564.2	8.6	21706.2	9.9
Eastern Europe								
High-Technology	414.0		2194.2		1737.5		1486.7	
Manuf. Goods	2758.7	15.0	14138.5	123.7	10925.3	15.4	9273.9	16.8
Total	3522.7	11.8	19460.9	11.3	16087.9	10.8	12425.7	12.0
Yugoslavia								
High-Technology	219.6		1181.0		952.9		782.5	
Manuf. Goods	1655.1	13.3	6813.3	17.3	5642.7	16.7	4672.9	16.8
Total	1871.7	11.7	7931.3	14.9	6679.5	7.3	784.9	8.0
Cuba								
High-Technology	18.7		77.6		87.1		47.0	
Manuf. Goods	261.2	7.2	698.7	11.1	692.6	12.6	361.8	13.0
Total	332.9	5.6	1302.7	6.0	1191.7	7.3	784.9	6.0

I.W. Exports to:	1970 $	% of	1980 $	% of	1981 $	% of	1982 $	% of
P.R.C.								
High-Technology	104.3		1073.4		1039.9		781.4	
Manuf. Goods	1062.9	9.8	8905.0	12.1	8043.5	129	6643.5	11.8
Total	1232.6	8.5	12440.2	8.6	11913.0	8.7	9760.1	8.0
Total All Comm. Countries								
High-Technology	1172.4		6934.2		5649.5		5315.8	
Manuf. Goods	8009.5	14.6	46144.3	15.0	39916.1	14.2	37136.8	14.3
Total	9521.9	12.3	61517.9	11.3	56990.7	9.9	50858.3	10.5
World								
High-Technology	24770.9		136205.3		140790.4		144788.0	
Manuf. Goods	162940.1	15.2	892324.8	15.3	867224.6	16.2	825540.9	17.5
Total	211644.5	11.7	1173144.0	11.6	1150371.2	12.2	1093046.8	13.3

Note: 17 industrialized Western countries included (I.W.) are: US, Canada, Japan, Belgium, Luxembourg, France, FRG, Italy, Netherlands, Austria, Norway, Sweden, Switzerland, U.K., Denmark, Finland and Ireland; High-technology products as defined by ITA; Office of Trade and Investment Analysis

Source: J.A. Martens - Quantification of Western exports of high-technology products to communist countries, Office of Trade and Investment Analysis, US Dept. of Commerce, Trade Research Division, Project No. DTR-26-83, April 1984, p. 7

Table 3.3. USSR Sources of I.W. High-Technology Products (millions of dollars)

	1970		1975		1980		1981		1982	
	High-Tech Exports to USSR	As % of Total	High-Tech Exports to USSR	As % of Total	High-Tech Exports to USSR	As % of Total	High-Tech Exports to USSR	As % of Total	High-Tech Exports to USSR	As % of Total
Canada	0.2	–	14.4	0.9	27.5	1.2	0.4	0.0	0.4	0.0
USA	12.5	3.1	219.2	13.5	84.7	3.6	56.5	3.2	32.1	1.5
Japan	43.5	10.8	169.2	10.4	400.2	17.2	366.0	20.0	378.2	17.6
Bel-Lux	5.9	1.5	26.5	1.6	18.0	0.8	12.1	0.7	15.4	0.7
Denmark	4.8	1.2	13.4	0.8	23.1	0.1	17.9	1.0	21.8	1.0
France	58.5	14.5	223.7	13.8	341.3	14.8	204.7	11.5	191.6	8.9
FRG	92.9	23.0	519.8	31.9	727.2	31.6	501.8	28.3	563.9	26.3
Ireland	–	–	0.3	–	0.2	Negl.	0.0	0.0	0.3	0.0
Italy	69.6	17.3	155.7	9.6	222.2	9.6	156.3	8.8	233.8	10.9
Netherlands	1.1	0.5	20.6	1.3	6.1	0.3	10.0	0.6	8.7	0.4
U.K.	56.0	13.9	51.7	3.2	125.7	5.5	132.5	7.5	96.9	4.5
Austria	5.6	1.4	27.4	1.7	48.2	2.1	30.4	7.7	47.4	2.2
Finland	6.3	1.6	35.9	2.2	86.2	3.7	121.8	6.9	375.6	17.5
Norway	0.1	–	13.4	0.8	12.3	0.5	6.5	0.4	9.6	0.4
Sweden	22.3	5.5	51.3	3.2	71.1	3.1	77.3	4.4	63.7	3.0
Switzerland	23.6	5.9	83.0	5.1	136.4	5.9	80.0	4.9	106.5	5.0
Total	402.9	100.0	1626.0	100.0	2330.0	100.0	1774.4	100.0	2145.7	100.0

Source: as in Table 3.2.

the third registered for Switzerland (!). In 1982, the US share in OECD exports of high-tech products to the Soviet Union was below that of Austria. Interestingly enough the Americans appear to have been the most restrictive in this regard with the Soviet Union (see Table 3.4.). On the other hand, Hungary, Rumania and Yugoslavia seem to be among the US favorites.

The significance of embargo measures are also better understood when specific high-tech items are taken into account, for example computer equipment, telecommunications equipment, measuring devices, etc. Although they frequently make up only a fraction of the value of some larger systems, they are of critical significance for their operation.

According to a recent OECD study on the research and development intensity of the OECD embodied technology export to CMEA countries, on an average, throughout the 1970s, only 2.6% of the total were high R+D intensive products, whereas over 77% were low R+D intensive goods. Among those countries importing the highest share of R+D intensive products from the OECD were Rumania and Hungary, with the Soviet Union beging clearly at the bottom of the ranking list (17).

Thus, the data referred to above indicate that when more closely analyzed, the embargo effects are not as insignificant and harmless as some authors claim. Its effectiveness, however, does not satisfy the US Administration. Therefore, since the beginning of the 1980s, it has undertaken several steps to improve its performance (18). These include inter alia:

1) Passing the Export Administration Amendments Act - EAAA in 1985, which provides the Defence Department with a larger role in export control. This is reflected in the adaptation of the Export Control List to the Military Critical Technologies List, developed by the Defence Department, the setting up of a National Security Control Office within the Defence Department, as well as in the introduction of an import ban for all companies that injure US export control.

2) Consolidation of the COCOM activities through the revision of the Export Control List, limiting the high-tech exports to the CMEA countries. It is supported by a newly established special military advisory group within the COCOM structure.

3) Intensified criticism and control of non-COCOM members which resulted in the introduction of more severe export control procedures, among others, by Austria, Switzerland and Sweden.

Table 3.4. Repartition of the Intentional Domination Effect of the US as Demonstrated by US High-Technology Exports to the Communist Countries, 1982 (millions of US dollars)

	Total Exports	Manuf. Exports	High-Tech. Exports	High-Tech. Exports as % of Total	High-Tech. Exports as % of Manuf.
Cuba	1.0	0.9	Neglig.	3.7	3.9
PRC	2904.5	1066.7	165.5	5.7	15.5
Yugoslavia	490.0	200.4	55.3	11.3	27.6
Bulgaria	106.5	35.4	6.7	6.3	18.9
Czechoslovakia	83.6	20.5	5.7	6.8	27.7
GDR	222.6	13.9	6.5	2.9	46.7
Hungary	67.8	58.9	13.8	20.4	23.5
Poland	292.6	65.5	8.2	2.8	12.6
Rumania	223.2	51.6	20.6	9.2	39.8
USSR	2589.0	598.4	32.1	1.2	5.4
Total Communist Countries	6997.5	2112.5	314.4	4.5	14.9
World	206044.7	140323.2	38461,3	18.7	27.4

Source: as in Table 3.2.

The same is true with respect to newly industrialized developing countries (NICs).

4) Allocation of additional funds to the US customs service for controlling the illegal transfer of US technology abroad.

As a result of these new steps, some companies deliberately gave up the idea of exporting more sophisticated goods to the East, as documented by a recent Philips statement (19).

A good example of the intensified attempts of the US Administration to control the technology export to the socialist countries is the recent case of the export contract of Toshiba Machine, covering the delivery of some special-purpose machine tools to the Soviet Union. To punish the company, which apparently injured COCOM regulations in this area, the US Administration imposed a 2 years' export ban on the company for the American market and demanded additional actions from the Japanese government. The Japanese government, in turn, punished the company by imposing a one year trade ban on its trade with CMEA countries. Also, the large Japanese trading house, C. Itoh, involved in the contract, was punished with a 3 month trade ban on its relations with CMEA countries. As a result, the two companies have announced their decision to withdraw from their initial plans to intensify the trade relations with CMEA countries (20).

The strategic export control measures are, however, not the only instrument aimed at the preservation of the dominant position of the West and hindering the East-West technology transfer. Another is connected to the measures leading to an increase in the transactional costs of the transfer operations. This includes inter alia bureaucratic obstacles in the West with respect to obtaining licenses for the export to the East and internationally coordinated restrictions on credit facilities. The degree of bureaucracy involved may be judged from the facts reported in Frankfurter Allgemeine Zeitung of January 22, 1985, according to which, from 1980-84 West German companies have completed 70 thousand applications for obtaining export licences to the Eastern countries. The companies complained about the complicated procedures involved, the unprecise formulation of the relevant laws and the resulting time losses and cost increases (21). The same is probably true with respect to technology exports for which COCOM approvals are necessary, as their number in the 1970s has increased substantially (Table 3.5.).

Table 3.5. COCOM Approval Requests

Year	Total Submissions	U.S. Submissions	U.S. Share (percent)
1962	124	2	1.6
1968	228	29	12.7
1970	544	133	25.6
1971	635	186	29.3
1972	1085	415	38.2
1973	1268	519	41.0
1974	1369	576	41.4
1975	1798	798	44.4
1976	1039	593	57.1
1977	1044	539	51.6
1978	1680	1050	62.5

Source: US Congress: Transfer of Technology to the Soviet Block, Hearing Lavvy Brady's Statement. Washington, GPO, 1980, p. 72

Similar cost-increasing effects resulted from the OECD International Export Credit Arrangement limitations on the maximum credit duration for the Eastern countries and the minimum value of the requested interest rate.

It may be quite instructive in this context to look at the sources of Western credits ensured for the CMEA countries as of the end of the 1970s. From the total value 24% were offered by West Germany, approximately 22% by France, approx. 16% by Japan, approx. 12% by Italy, 6% by Austria and only around 3% by the US (22).

Last but not least, a significant instrument in achieving the domination stoppage effect applied by the West in its relations with the East is associated with the limitation of the access to Western educational and scientific communication systems (23). As reported by H.D. Jacobsen, this is not limited solely to CMEA countries but includes all non-American citizens (24). Its degree and effectiveness is hard to assess, due both to the specific nature of the activities involved and to the intertwining of the Western and Eastern policies involved. The final effect, however, is a low degree of penetration of the Western educational and research facilities by Eastern bloc citizens (see Table 3.6.).

Table 3.6. US-USSR Scientific Personnel Exchange under the Bilateral Agreements

| | Participants in Bilateral Agreements | | |
Year	US to Soviet Union	Soviet Union to US	Total
1972	262	246	508
1973	404	273	677
1974	837	844	1681
1975	1170	1114	2284
1976	983	853	1836
1977	761	664	1425

Source: US Congress: Key Issues in US-USSR Scientific Exchanges and Technology Transfer Hearings. Washington, GPO, 1979, p. 128

If judged by the US-Soviet scientific personnel exchange, the personnel flows seem to be practically insignificant and far below the levels recorded among the Western countries themselves or among the Western and developing countries. The figures quoted in Table 3.6. point out that a strict rule of reciprocity is apparently applied by the two partners, thus indicating that no special allowances are considered.

5. Conclusions

As demonstrated by the preceding analysis, East-West technology transfer occurs under the conditions of the dominant position of the Western economies and, as a result, the domination effect is exerted on the economies of the East. This domination effect is both of an unintentional and an intentional nature and is not limited to technological relations but covers a much larger area of economic activities.

The overall technological lead of the OECD countries over the socialist bloc compel the latter to apply a strategy of technological catching up and structural adjustment. Attempting to both rationalize and accelerate the process of technological catching up, the CMEA countries supplement their internal efforts with direct technology imports from the West. Here, however, due to the basic differences in the political and ideological aims and values between themselves and the Western countries, they are confronted with the intentional domination effect of the latter, aimed clearly at the preservation of the technological distance separating them from the socialist countries. The measures applied include embargos, some elements of economic warfare, strict credit and financial terms, restrictions on the access to non-commercial scientific and technical information as well as hindrances to market access. All of these must influence the international technology transfer performance of the socialist countries and must inevitably be taken into account in any comparisons with other technological importers. All of these elements must also be taken into account by the Eastern policy-makers in their decisions related to national and regional technological policies.

The harder the access to Western technology and the higher the costs of its transfer, the more rewarding chances will be sought in intra-regional cooperation schemes and rearrangements of national systemic conditions and policies towards improving these countries' innovative performance.

References and Footnotes

1   K.W. Rothschild (ed.) - Power in economics. Selected readings, Penguin Books, 1971, p. 7

2   F. Perroux - Une théorie de l'économie dominante, Economie Appliquée, Archives de l'I.S.E.A., April-September 1948, nos. 2-3 (2); F. Perroux - Le dynamisme de la domination, Economie Appliquée no. 2, 1950; F. Perroux - La concurrence et l'effet de domination, Banque, May 1952

3   F. Perroux - The domination effect and modern economic theory, Social research, Vol. 17, 1950, p. 186

4   F. Perroux - Wirtschaft und Macht, Verlag Paul Haupt, Bern und Stuttgart, 1983, p. 14

5   F. Perroux - Wirtschaft ..., op.cit., p. 46

6   F. Perroux - Wirtschaft ..., op.cit., p. 54

7   F. Perroux - L'Economie du XXième siècle, Press Universitaires de France, 1961, p. 28

8   F. Perroux - L'Economie du XXième siècle, op.cit., p. 31

9   F. Perroux - L'Economie du XXième siècle, op.cit., p. 46 and p. 93

10 F. Perroux - L'Economie du XXième siècle, op.cit., p. 93

11 F. Perroux - L'Economie du XXième siècle, op.cit., p. 94

12 F. Perroux - Wirtschaft ..., op.cit., p. 50

13 Recent decisions of the Soviet Union to establish some initial contacts with GATT and its apparent desire to join IMF may be considered as the final steps in the process described

14 J.P. Hardt, K. Tomlison - Potential role of Western policy towards Eastern Europe in East-West trade, in A. Becker (ed.) - Economic relations with the USSR in the 1980s, Lexington Books, 1983

15 P. Hanson - Western economic sanctions against the USSR: their nature and effectiveness, in: External economic relations of CMEA countries: their significance and impact in a global perspective, NATO, April 1983, p. 71

16 P. Hanson - Western economic sanctions ... op.cit., p. 84

17 H. Wienert, J. Slater - East-West technology transfer. The trade and economic aspects, OECD, Paris 1986, p. 260

18 H.D. Jacobsen - Die Technologiekontrollpolitik der Vereinigten Staaten und ihre Auswirkungen auf die West-West Beziehungen, Europa Archiv, Nr. 15, 1986, pp. 443-450

19 H.D. Jacobsen - Die Technologiekontrollpolitik ... op.cit., p. 449

20 Rynki Zagraniczne, nr 93 (4713), 5.08.1987, p. 1

21 Die Cocom-Regeln werden präzisiert, Frankfurter Allgemeine Zeitung, no. 18, 22.01.1985

22 Z. Baka - Wojna ekonomiczna, Wydawnictwa MON, Warszawa
    1983, p. 161
23 Scientific communication and national security, National
    Academy of Sciences, Washington D.C., National Academy
    Press, 1982
24 H.D. Jacobsen - Die Technologiekontrollpolitik ... op.cit., p.
    448

Part II
Quantifying East-West Technology Transfer

Chapter 4
World-Wide Transborder Technology Flows

Dominant Centers and Dependents

1. Introduction

Measuring transborder technology flows is by all standards a complex undertaking. This is both a result of the substance of the phonomenon as well as the state of the relevant statistics. With regard to the first issue, it is principally the nature of the technology which makes its strict statistical delineation nearly impossible (1). Technology is a part of human knowledge - and, so far, we are unable to measure knowledge in a quantitative way. This knowledge may be transmitted by a variety of means, including people, products, documents and, transfer of information via telecommunication facilities. Some of these flows will be reflected in commercial transactions while others definitely will not. Hence, even if comprehensive statistics were compiled and thereafter made available, it would still not allow us to calculate precisely the size, structure and importance of the transborder technology flows. It would, however, have undoubtedly helped us. This is unfortunately not the case and the required statistics are basically lacking. Only recently some attempts to alter the existing situation have been undertaken. Here, one should specifically mention the efforts of the OECD to buildup technology transfer data bank and the measure taken by the Economic Commission for Europe (ECE), to set up a data bank for intra-ECE technology transactions (2).

In effect, when discussing transborder technology flows, some proxy measures must be used. These include, as a rule, trade in machinery and equipment, foreign direct investment flows and last but not last, licensing payments (3). In the following, I shall use all three forementioned proxies to review the current distribution of dependency and domination in the world of technology.

2. The World of Technology. Global View

To better understand the situation in international technology flows, we shall start our analysis with a brief discussion of the world-wide distribution of technological potential. This potential is primarily associated with research and development - currently the major source of technological advances. It is estimated that the world spent over US $ 295 billion on R and D in 1983, and thus more than doubled the level of outlays registered in 1970. From the total amount, $ 192 billion were spent by developed countries (DMECs), $ 71 billion by the socialist countries of Eastern Europe and the rest by the developing world (See Table 4.1.).

Table 4.1. World-wide Distribution of R and D Expenditures (in percentages of world total)

Regions/Countries	1970	1975	1980	1983
1. World total	100.0	100.0	100.0	100.0
2. DMECs there from	77.5	70.2	72.7	69.0
EEC	20.3	21.6	21.5	20.9
Japan	6.7	9.5	11.7	12.6
United States	39.9	33.3	33.7	33.4
3. Socialist countries of Eastern Europe (SCEE) there from	25.2	27.1	24.4	24.0
USSR	19.4	20.9	17.6	18.0
4. Developing countries	2.3	2.7	2.9	7.0

Source: OECD/STIIU Data Bank, November 1985. Estimates of the CMEA secretariat and own estimations

Among the developed countries, the United States and Japan accounted for about 67% of the total.

In the case of the socialist countries, over 74% of the total R and D outlays are born by the Soviet Union alone. This means that the three countries together control the bulk of world-wide technological resources.

As far as the number of scientists and engineers engaged in R and D is concerned, the distribution world-wide departs significantly from that of R and D expenditures. It is estimated that around 60% of the world total R and D personnel is employed in socialist countries, around a third of the world total in developed countries, and the remainder in the developing world. On the whole, the world total employment of scientists and engineers come close to 3 million.

If we look at the process of transborder technology flows from a long-term perspective, we can see that within the last twenty or so years a dramatic upsurge in the growth of technology-related flows occured. This was particularly pronounced in the 1960s and early 1970s, which can be attributed both to the overall economic prosperity during this period and to the US's desire to exploit their technological superiority in the expanding EEC market (4).

Since the mid 1970s, a clear slowdown in technology flows has occured which was basically the result of the deterioriation of the overall economic climate and the structural crisis initiated by the oil shock. Still, however, technology related flows continued to expand, although at a lower pace (see Table 4.2.).

The period since the beginning of the 1980s has witnessed a drastic slowdown in most of the technology-related flows.

Particularly hard hit were the developing countries and, to a lesser extent, the socialist countries. In both cases, the eruption of the debt crisis and resultant poor financial liquidity are to blame.

The current situation in worldwide technology flows is characterized by a clear concentration of these flows in the group of industrialized countries, memebers of the Organisation for Economic Cooperation and Development, further referred to as OECD countries.

This group includes both major exporters as well as majors recipients of the technologies transferred internationally (see Table 4.3.). These countries represent over four-fifths of the international supply of machinery and transport equipment; they provide

Table 4.2. Indicators of International Technology Flows, 1962-1985
(Billions of dollars, Current prices)

Indicator	1962	1972	1973	1975	1976	1979	1981	1982	1983	1984	1985
Capital goods trade (a)											
DMECs:											
1. Exports to world	26.6	85.8	113.4	176.4	195.7	310.0	349.8	337.6	328.6	357.4	375.6
2. Imports from world	13.7	60.4	80.8	107.7	122.4	204.7	231.0	227.6	236.7	276.5	298.9
3. Exports to LDCs	6.5	20.9	27.8	58.0	66.2	92.6	123.6	116.9	102.0	102.2	96.4
4. Imports from LDCs	0.1	1.2	2.3	3.6	4.8	10.4	15.0	16.5	21.1	28.1	28.5
Foreign direct investment											
5. Flows from DMECs to world of which to:	4.7	14.5	24.0	27.2	27.0	58.3	50.6	23.4	31.8[b]	40.0[b]	57.2[b]
6. Other DMECs	2.8	9.2	13.4	14.5	13.7	32.8	42.2	30.1	33.6	39.9	33.7
7. Flows from DMECs to LDCs (c)	1.4	4.4	6.7	10.5	7.8	13.5	15.3	10.4	7.8	11.3	7.7
Receipts of royalties and fees											
8. Flows to DMECs	-	3.9	4.5	5.9	6.4	8.8	11.1	11.2	12.1	-	
9. Flows from LDCs (d)	-	0.7	0.7	1.0	1.0	1.7	2.1	2.0	2.3	2.2	2.3
Technical assistance											
10. Flows to LDCs	0.7	1.8	2.3	2.2	2.9	4.7	5.2	5.4	5.8	5.9	6.0

a SITC 7 - (7173 + 7194 + 7247 + 7242 + 7250 + 7291 + 7296 + 7321 + 326 + 7329 + 7331)

b Including transactions between United States parent companies and their financial affiliates in the Netherlands Antilles

c These data are not fully comparable with those shown for indicators 5 and 6

d Flows to the Federal Republic of Germany, Japan and United States only

DMECs - developed market economy countries, LDCs - developing countries

Source: Trade and Development Report, 1987, UN, New York 1987, p. 88

Table 4.3. Transborder Technology Flows. Selected Indicators (in % of the world total)

	Exports			Imports		
	1970	1975	1982	1970	1975	1982
1. Machinery and transport equipment	100.0	100.0	100.0	100.0	100.0	100.0
OECD	87.5	87.0	84.6	66.2	57.7	58.6
LDCs	1.6	2.9	6.3	21.5	28.9	30.9
CMEA	10.7	10.0	8.9	10.5	11.5	8.7
2. FDI flows	100.0	100.0	100.0	100.0	100.0	100.0
OECD	-	99.1(a)	96.4	-	58.4(a)	61.3
LDCs	-	0.9(a)	3.6	-	41.6(a)	38.7
CMEA	-	-	-	-	-	-
3. Licensing payments (fees and royalties)						
OECD	98.5	98.5	99.0(b)	80.0	83.0	85.0
LDCs	1.0	0.8	0.5(b)	10.0	9.0	10.0
CMEA	0.5	0.7	0.5(b)	10.0	8.0	5.0

a - 1976   b - 1981

Source: Computed from the Handbook of International Trade and Development Statistics, Supplement 1985, UN, New York 1985; K.H. Oppenländer - Auslandsinvestitionen und außenwirtschaftlicher Technologietransfer: Übersicht über Ursachen und Ausmaß, in: L. Späth, H. Dräger (eds.) - Zielsetzung Partnerschaft. Die weltwirtschaftliche Bedeutung von Auslandsinvestitionen und Technologietransfer, Band 9, Bonn Aktuell, Stuttgart 1985, p. 60; J. Maciejewicz - Międzynarodowa konkurencja technologiczna, Sprawy Międzynarodowe 3/1986, p. 41

most of the world's foreign direct investment flows and receive the majority of the world's fees and royalties. The OECD produced technologies thus constitute today the core of the world technological system. As fas as the import side is concerned, these proportions are somewhat different, still, however, the bulk of worldwide technology transactions are associated with the OECD group. The leading position of the OECD countries seems to be stable and in the course of the 1970s it was only slightly weakened.

The position of the two other major actors in the international economic scene - the less developed countries (LDCs) and the socialist countries of Eastern Europy (CMEA) - is similar to a considerable degree. Both of them represent marginal shares of the international technological supply and quite substantial shares of the world demand. In the course of the 1970s some important changes in the mutual power relationship of the two tookplace, with respect to technology flows. There has been a substantial improvement in the LDCs' position, both as technology exporters and recipients and a slow but steady erosion of the position held by CMEA countries, again both as technology exporters and recipients.

A peculiar feature of the current distribution of the transborder technology flows is a high degree of their regional endogenization (internalization). As indicated in Table 4.4., in the 1970s, an average of around two-thirds of OECD and over 70% of the CMEA machinery exports were of an intra-regional nature. Only LDC countries followed a different pattern. Here, the predominant part of their import of machinery, as well as over 50% of their export of machinery, took place in OECD countries. Thus, one could assume that the OECD region and LDC countries are largely operating within the framework of the same technological base - a base which is of overwhelming OECD origin. The CMEA group, on the other hand, seems to be mainly oriented towards its own technological resources.

The degree of endogenization of transborder technology flows, diluted somewhat by the behavioral patterns of the LDCs in the machinery trade, is apparently much stronger in the case of foreign direct investments and licensing operations. It is estimated, for example, that up to 95% of the total foreign direct investments of those countries is located in other developing countries (5). The same is true with respect to licensing operations and the trade in technical services (6). Thus, in general, the current-

Table 4.4. Endogenization of the Transborder Technology Flows. Trade in Machinery and Transport Equipment (Share of intra-regional trade)

	World		OECD		LDCs		CMEA	
	1970	1982	1970	1982	1970	1982	1970	1982
Exports to								
OECD	100.0	100.0	73.8	64.4	21.8	31.5	3.0	2.7
LDCs	100.0	100.0	52.9	54.2	44.2	42.2	0.3	0.3
CMEA	100.0	100.0	6.5	7.6	15.4	16.6	73.0	71.5
Imports from								
OECD	100.0	100.0	96.9	93.0	1.3	5.8	1.1	1.2
LDCs	100.0	100.0	88.7	86.1	3.3	8.6	7.7	4.8
CMEA	100.0	100.0	25.2	26.4	0.05	0.2	74.5	73.2

Source: Computed from the Handbook of International Trade and Development Statistics, Supplement 1985, UN, New York 1985, p. A 38

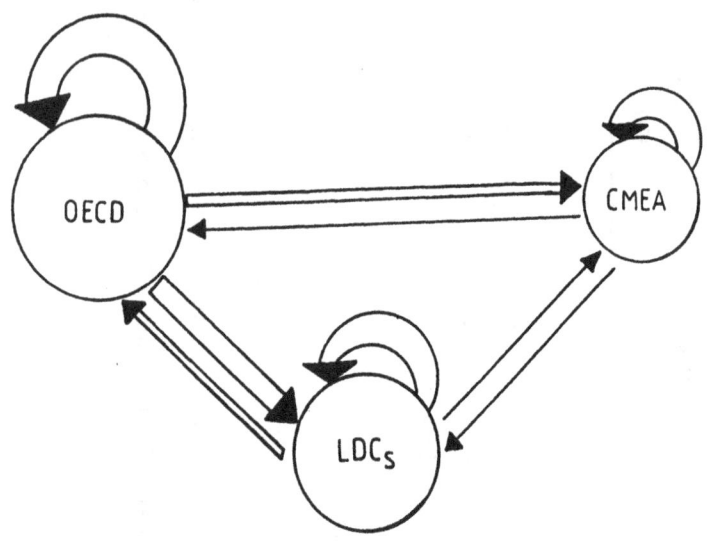

Fig. 4.1. Major Directions of Transborder Technology Flows

situation in transborder technology flows may be illustrated as in Fig. 4.1.

3. The World of Technology. A Disaggregated View

Let us now depart form these general observations and look more closely into the repartition of the transborder technoloy flows within the three identified groups of countries.

As fas as the trade in machinery and transport equipment is concerned, approximately 50% of the OECD exports in the 1970s was made up by EEC member countries, followed by the US (ca. 20%) and Japan (ca. 15%). These proportions were fairly constant with the exception of Japan, which increased its share from 10% of the OECD total in 1970 to 18.1% in 1982 (see Table 4.5.). On the import side, EEC countries are again in the top (ca. 45% of the OECD total), followed by the US (around 20%). The share of

Japan is negligible, accounting for around 3% of the OECD total. Hence, it turns out that the major streams of technology flows via machinery and transport equipment within OECD area are concentrated in Western Europe and the United States, accounting together for around two-thirds of the OECD total.

As far as LDC countries are concerned, major exporters of machinery and transport equipment are coming from South and South-East Asia (so called newly industrialized countries - NICs), which accounted for about 70% of the total LDCs export in this category in the 1970s. Interesting is the fact that these countries were able to increase their share in LDCs machinery export from 63.8% in 1970 to 74.1% in 1982. Among these one should specifically mention such large exporters as Singapore, South Korea, Hong Kong and India.

The remaining part of LDCs machinery exports is accounted for principally by newly industrialized South American countries (Brazil and Argentina) whose share, however, has decreased from 26.7% of the LDCs total in 1970, to 16.5% in 1982, and by West Asia countries, representing ca. 7% of the LDCs total. The import of machinery and equipment by LDCs has a very different geographic structure and is more or less equally divided among the three identified subgroups of countries, each of them accounting for around 30% of LDCs total in the 1970s.

Hence, to summarize, the transborder flows of machinery and equipement in the case of LDC countries seem to be highly concentrated on the export side in South and South East Asia, whereas a quite balanced picture prevails as far as imports are concerned. Newly industrialized countries are constantly increasing their share in the LDCs total.

Now, if we look at the CMEA member countries, we note quite a peculiar structure of their transborder flows of machinery and equipment. Namely, in imports, the Soviet Union clearly dominates, accounting for ca. 50% of the CMEA total, whereas in exports, the share of the Soviet Union is ca. 1/4 of the CMEA total.

All in all, the foregoing discussion points out that in the case of transborder flows of machinery and transport equipment, individual regions have their clear dominant centers and their dependents, adding to the complexity of the domination and dependency existing at the inter-regional level. Interestingly enough, export sources are far more concentrated than import sources. This observation holds also its validity in the case of foreign direct in-

Table 4.5. World Trade in Machinery and Transport Equipment Broken Down by Principal Regions and Countries (millions of US dollars, fob)

Export	Import Year	OECD				LCDs		CMEA		World
		EEC	USA	Japan	America	West Asia	South East Asia	Total	Soviet Union	
OECD	1970	25504	11080	1557	6262	2020	4552	2381	1122	78623
	1975	61435	22380	2787	17595	156673	13359	9855	4840	212665
	1982	129709	60157	6513	31241	39487	38417	11707	7479	430860
EEC	1970	17535	3676	491	2203	1225	1466	1493	614	40286
	1975	42565	7219	843	5433	9143	3842	5951	2666	107447
	1982	86050	15495	1649	7263	18972	10185	5149	2766	198431
USA	1970	4132	–	1146	2847	4411	1098	69	45	17822
	1975	7780	–	1698	7490	4022	3519	767	547	45710
	1982	18644	–	4433	14232	7597	11356	354	229	90249
Japan	1980	977	2423	–	479	175	1674	126	103	7834
	1982	3131	5972	–	2540	2142	4991	835	558	27405
LDCs	1970	150	518	43	198	69	273	5	2	1431
	1975	925	1767	345	1107	615	1257	43	25	7119
	1982	4487	9797	1123	2574	2955	5992	87	72	31228
America	1970	38	146	–	183	–	3	–	–	382
	1975	139	288	49	1032	24	26	7	1	1778
	1982	886	1460	113	1815	197	196	2	–	5289

Export	Year	OECD			LCDs			CMEA		World
		EEC	USA	Japan	America	West Asia	South East Asia	Total	Soviet Union	
West Asia	1970	18	2	–	–	41	2	–	–	89
	1975	90	23	–	1	363	8	–	1	613
	1982	259	49	11	3	1896	183	2	1	2710
South East Asia	1970	75	368	43	15	22	268	3	1	898
	1975	629	1447	295	74	214	1221	26	20	4529
	1982	3201	8279	999	751	848	5611	79	70	23804
CMEA	1970	292	12	16	369	451	251	7040	3125	9647
	1975	1013	54	14	795	1287	416	18071	7780	24530
	1982	1596	154	50	1957	2533	903	32483	14446	45406
Soviet Union	1970	60	–	4	238	284	132	1428	–	2775
	1975	167	6	4	474	613	197	3709	–	6378
	1982	271	9	7	1375	811	606	5482	–	11368
World	1970	25946	11610	1817	6835	2541	5103	9444	4258	89769
	1975	63380	24202	3147	19508	18582	15079	28005	12660	244443
	1982	135825	70165	7692	35808	45241	45781	44352	25048	509385

Source: Handbook of International Trade and Development Statistics, Supplement 1985, UN, New York 1985, pp. A.38–A.40

Table 4.6. Value of Foreign Direct Investments of the Leading Western Economies

Country	Value of the Investments Abroad		Value of the Foreign Investments at Home		Remarks
	1976	1980	1976	1980	
USA ($ billions) (1)	136.809	192.648	30.770	54.462	Book value. Data of the US Department of Commerce (1) Based on 1966 Census (2) Based on 1977 Census
(2)		313.468		65.483	
UK (£ billions)	22.265	31.730	16.000	24.725	Book value. Bank of England estimates
Switzerland (SFRS bill.)	45.5	58.5	11.5	15.0	Estimate of the Schweizerische Bankgesellschaft
FRG (DM billions) (1)	47.048	74.162	45.518	57.616	(1) Cumulative net transfers (2) Assets statistics Data of the Deutsche Bundesbank
(2)	48.377	83.334	78.899	93.771	
Japan ($ billions)	19.405	31.804	1.695	2.679	Cumulative annual permissions Ministry of Finance
France (FF billions)	38.5	77.8	48.3	93.7	Banque de France Cumulative net capital flows
Canada (Can $ bill.)	11.501	13.443	43.335	46.951	Book value. Ministry of Industry, Trade and Commerce
Italy (Lit, billions)	2.491	6.846	5.035	8.274	Banca d'Italia. Assets value

Source: D. Kebschull– Auslandsinvestitionen und Technologietransfer aus der Sicht der kapitalexportierenden Industrieländer – Auswirkungen und Förderungsmaßnahmen, in: L. Späth, H. Dräger (eds.), op.cit., pp. 168-169

Table 4.7. World Trade in Licenses 1970-83
(% of the world total)

Regions and Countries	1970	1975	1981	1983
Exports				
OECD	98.5	98.5	99.0	99.0
USA	58.3	57.6	52.0	58.8
U.K.	8.5	8.1	8.1	8.8
Switzerland	6.0	8.4	7.9	-
Italy	2.0	4.3	6.1	1.6
FRG	3.2	4.2	3.9	5.0
France	1.7	2.5	3.5	7.6
Japan	1.4	1.9	3.4	8.8
Netherlands	2.5	2.5	2.8	2.8
LDCs	1.0	0.8	0.5	0.5
CMEA	0.5	0.7	0.5	0.5
Imports				
OECD	80.0	82.0	85.0	88.5
Japan	10.3	9.2	12.2	14.4
Italy	7.8	9.7	8.8	9.3
FRG	8.6	11.0	8.4	14.0
France	5.0	6.9	6.8	11.8
U.K.	6.3	7.1	6.3	9.9
USA	5.6	6.3	5.1	2.5
Netherlands	2.9	3.7	4.3	6.7
LDCs	12.0	8.0	10.0	8.0
CMEA	8.0	10.0	5.0	3.5

Source: J. Maciejewicz - Miedzynarodowa konkurencja technolog-iczna, Sprawy Miedzynarodowe 3/1986, p. 41; OECD Science and Technology Indicators, No. 2, OECD, Paris 1986, p. 55

vestments, where more than 50% of the total cumulated book value can be attributed to US investors, followed by the U.K. and Switzerland (see Table 4.6.).

As far as LDC countries are concerned, a bulk of the investments comes from South-East Asia (mainly Hong Kong, Singapore and the Philippines), followed by some Latin American countries (mainly Brazil, Argentina and Venezuela) (7). Here, the newly industrialized countries dominate.

With regard to the world trade in licenses, there is again a clear concetration of the flows in some countries. Thus, in the export of licenses, nearly 60% of the world total is accounted for by the United States, followed by the U.K. (over 8%) and Switzerland (see Table 4.7.). Large exporters of licenses also include Japan, France and West Germany.

As far as the import of licenses is concerned, the distribution among OECD countries in the 1970s was far less concentrated than in the case of exports, which is mainly due to the low share of the US. Accounting for over 50% of the world licensing income, the said country represents only ca. 5-6% of the world total licensing payments. The import of licenses to LDCs and CMEAs on the other hand, is largely insignificant and, what is more, it can be attributed to only a handful of the said countries: Brazil, Mexico and Argentina in LDCs and the Soviet Union, Poland and Hungary in CMEA.

The description of transborder technology flows would be incomplete without pointing out that to a high degree the current flows are internalized within transnational corporations (TNCs). This is especially true with resepct to intra-OECD and OECD-LDCs flows. It is much less so, as far as intra-LDCs flows are concerned, and obviously enough it is practically negligible in the case of CMEA countries. The precise degree of this internalisation is unknown but available indices point out that as much as over 50% of the technology flows remain within the same ownership system. In addition, it is also significant that this is apparently a better part of the technology traded internationally. The external transfers are believed to contain mostly second rate and outdated technological solutions (8).

4. Institutional Forms of Transborder Technology Flows

Technology may be transferred internationally through a variety of institutional arrangements (transactions), the combination of which is largely determined by the systemic characteristics of the parties involved and the structure of their motivations and interests. Putting aside all non-commercial forms of transborder technology flows, such as the exchange of books and journals, the movement of people, etc., the remaining arrangements may be grouped into the three distinct classes (9):

a. pure foreign trade transactions (sale of products and knowhow),
b. regulated foreign trade transactions (all forms of cooperation rrangements, such as buy-back agreements, compensationary transactions, sub-contracting, etc.),
c. pure foreign direct investments transactions.

Historically, the pure foreign trade transactions dominated the scene but in the 1960s these were largely substituted by the pure foreign direct investment transactions. Since the 1970s onwards, a substantial increase in the regulated foreign trade transactions has taken place (see Fig. 4.2.).

This has been well proved by a recent empirical research by the West German Institute for Economic Research (IFO). In a study conducted in 1983 covering a few hundred leading West German companies, it was discovered that in their relationships with other OECD countries, as well as particularly with LDCs countries, the share of the companies basing their foreign technological strategy on regulated forms of technology transfer is substantially increasing, whereas the share of those using exclusively direct investments, or mixed forms, is rapidly decreasing (10).

As far as the East-West technology flows are concerned, they were, and still are, characterized by the predominant use of the pure foreign trade transactions. In the 1970s, however, an increase in the regulated foreign trade transactions occurred. As argued by A. Wass von Czege, this can be largely explained by the apparently improved adaptation of these institutional forms of the specific properties of the technology transfer process, on the one hand, and to the specific systemic arrangements prevailing in the Eastern countries, on the other hand. At the same time, they also allowed for more control and coordination by the Western counterparts, although not as much as pure direct investment transac-

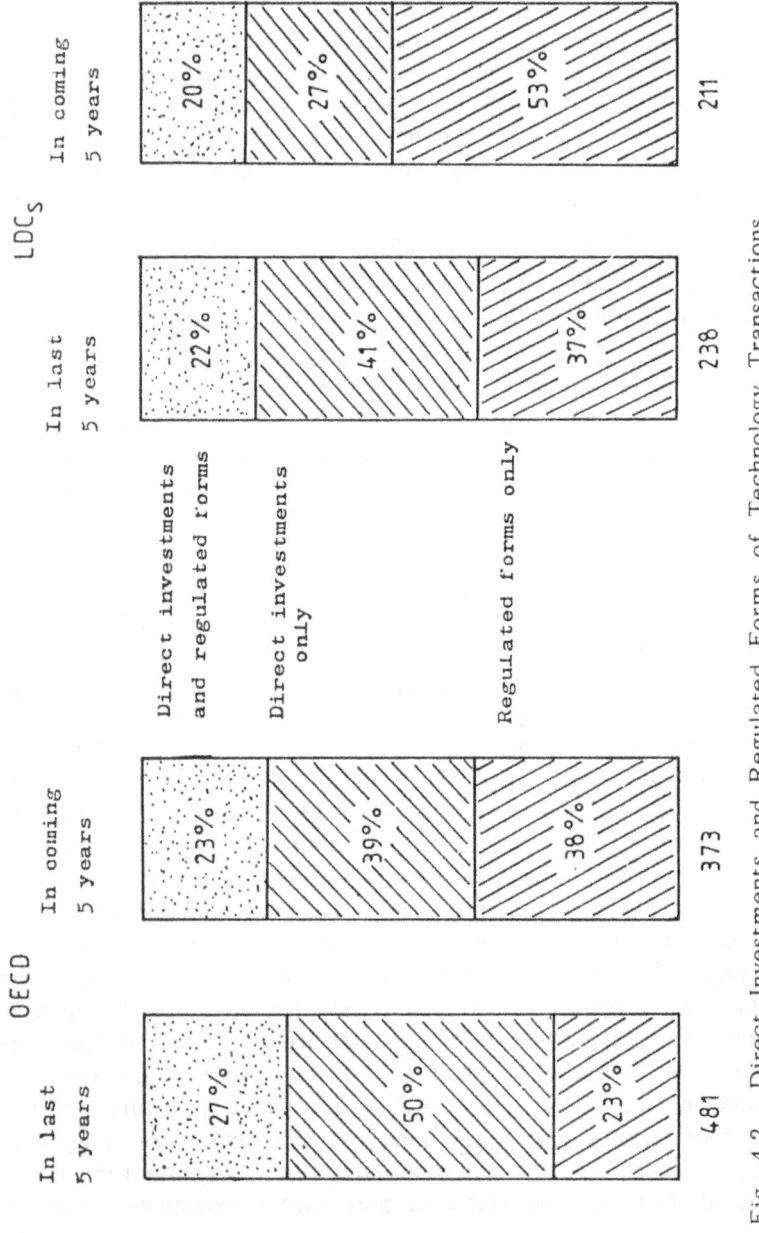

Fig. 4.2. Direct Investments and Regulated Forms of Technology Transactions

Source: K.H. Oppenländer – Auslandsinvestitionen ... op.cit., p. 54

tions (11). As such, they might be considered as a mutual compromise reached by both the Eastern and Western partners.

This compromise was subsequently followed by a decrease in foreign direct investment links, predominantly of minority foreign ownership, which again, might be taken as yet another compromise.

References and Footnotes

1   M. Simai - International technology transfer and economic development in the late 20th century, Trends in World Economy, No. 48, Hungarian Scientific Council for World Economy, Budapest 1984, p. 16

2   See ECE, Transfer of technology. Development of statistics: Requirements for the successful transfer of technology. SC.TECH/R. 182, 8 August 1985

3   J. de Castro, T. Ganiatsos, A. Olechowski, H. Qagaya - Changes in international economic relations in the last two decades, Trade and Development. An UNCTAD Review, No. 5/1984, pp. 27-36

4   J. de Castro, T. Ganiatsos, et al. - Changes in ... op.cit., p. 27

5   J. Monkiewicz - Multinational production enterprises: a preliminary overview, UNIDO/PC. 121, September 10, 1985, pp. 8-9

6   J. Monkiewicz - Technology exports from developing countries. Dimensions, nature, potentials and issues, UNIDO/IS. 525, March 28, 1985, pp. 7-9

7   L.T. Wells - Third world multinational, The MIT Press, 1983, p. 10

8   E. Mansfield, A. Romeo, S. Wagner - Foreign trade and US research and development, Review of Economics and Statistics, Vol. 61, 1979, pp. 49-57

9   See K.H. Oppenländer - Auslandsinvestitionen und ... op.cit., pp. 42-50

10 G. Pollak, J. Riedel - Das Engagement deutscher Unternehmen in Entwicklungsländern - Stand und Perspektiven, Ifo-Schnelldienst 21/1984, p. 21

11 A. Wass von Czege - Mechanismen zum intersystemaren Technologietransfer - ihre Klassifizierung und unterschiedliche Bewertung in Ost und West, IA u. U., Universität Hamburg, Forschungsbericht Nr. 11, Hamburg 1977, pp. 13-25

Chapter 5
Foreign Technology Infusion into CMEA Countries

Size, Structure, Sources

1. Introduction

CMEA countries have always considered foreign technology as secondary to the local technological effort, although its role has fluctuated considerably over time. It should also be pointed out that throughout the entire postwar period the technology import politics of the socialist countries were clearly intra-CMEA oriented and the bulk of the technology import of individual CMEA countries originated from within this grouping. The socialist countries acted as both the main technology suppliers as well as recipients. A relatively large import of Western technology in the 1970s did not change the overall picture. Still, the main share of the infused technology was of local origin. The change which occurred was related almost entirely to the import structure of the most modern technologies, in which case the predominant part was of Western origin.

Western technology has been infused into CMEA countries via pure trade arrangements, regulated foreign trade transactions as well as via foreign direct investments. Pure foreign trade transactions clearly dominated the scene, which is generally attributed to the policy priorities of the CMEA countries, apparently avoiding direct capital links with the West. These transactions were to be carried at in the form of various industrial cooperation agreements, which were given considerable economic and political attention in the first half of the 1970s. Industrial cooperation arrangements constitute a departure from the classical market coordination mechanism. Under new conditions, the coordination is provided via relevant agreements reached between the parties both at the microas well as the macro levels. Therefore, on can speak of regulated foreign trade transactions.

2. Pure Foreign Trade Transactions

The principal carriers of Western technology to the CMEA region throughout the 1970s and beginning of the 1980s were capital goods and technology - related intermediate inputs. The total value imported to CMEA increased from ca. $ 19 billion in 1975 to $ 26 billion in 1980, to decline thereafter to $ 22 billion in 1983. From the seven countries making the CMEA total, the Soviet Union alone accounted for 49% in 1975, 52% in 1980 and 65% in 1983 (1). The share of the above two items in overall OECD exports to CMEA has shown great variations over the course of time. In 1970, they accounted for around 2/3 of total OECD exports to CMEA, to increase to nearly 3/4 of the total by 1975. Since then, however, their share has declined to 61% of the total in 1983. The aforementioned shifts were mainly the result of the changing share of the capital goods, with the intermediate inputs retaining a fairly constant share: slightly above 1/3 of the total in 1970 and a little more in 1980 (2). It is interesting to note that Western technology exports to CMEA outpaced the overall growth rate of the East-West trade only in the first half of the 1970s and lagged significantly behind thereafter (see Table 5.1.). This means that the phase of rapid expansion of technology imports to CMEA burnt itself out quickly.

As far as capital goods are concerned, the single largest importer from the West was the Soviet Union, accounting for 50.8% of the total CMEA imports in 1970, and for nearly 72% in 1983. It was followed in the 1970s by Poland and Rumania, which were thereafter replaced by Hungary and Czechoslovakia. Similar developments have taken place in technology-related intermediate inputs.

To measure the relative intensity of the technology import to the individual CMEA countries, we can weigh their relevant shares in import with their share in the regional Net Material Product. The results of the calculations performed are presented in Table 5.2.

Table 5.1. Western Technology Exports to the East and Total East-West Trade (indices (1970 = 100) based on data in US dollar prices of 1970)

	1975	1980	1982	1983
A. <u>Western technology exports to the East</u>				
OECD capital goods exports to the East:				
Eastern Europe	209	137	107	104
Soviet Union	210	167	237	270
Total	120	153	173	188
OECD technology related intermediate inputs (exports to the East)				
Eastern Europe	183	189	130	135
Soviet Union	252	372	361	394
Total	208	257	215	230
B. <u>Total East-West trade</u>				
Total OECD exports to the East				
Eastern Europe	179	191	141	136
Soviet Union	265	346	428	462
Total	215	256	260	271
Total OECD imports from the East				
Eastern Europe	128	162	146	148
Soviet Union	126	147	156	157
Total	127	156	150	152

Source: OECD East-West technology transfer data base (Western statistics), quoted form H. Wienert, J. Slater - East-West technology transfer ..., op.cit., p. 396

Table 5.2. Western Technology Exports to the East and Technology Import Intensity (shares of Eastern countries in respective seven-country totals)

	Western capital goods export					Western intermediate (technology) goods export					Coefficient of the technology import intensity					
											Imports of capital goods			Imports of intermediate technology goods		
	1970	1975	1980	1982	1983	1970	1975	1980	1982	1983	1970	1975	1982	1970	1975	1982
Bulgaria	4.4	5.3	3.8	5.3	5.4	6.7	4.1	4.7	5.1	5.3	2.32	2.65	2.30	3.53	2.05	2.22
Czecho-slovakia	11.7	6.6	8.0	6.0	5.3	10.7	7.3	6.9	6.7	7.1	2.34	1.37	1.30	2.14	1.52	1.45
German Democratic Repulic	7.0	3.4	4.8	5.3	6.0	6.0	4.6	4.7	3.5	4.7	1.15	0.57	0.88	1.00	0.77	0.58
Hungary	5.4	4.2	6.9	7.0	5.4	12.2	9.5	10.1	11.6	11.1	2.00	1.41	2.70	4.52	3.60	4.46
Poland	9.6	22.6	14.8	4.6	4.7	15.0	19.7	13.6	8.6	8.9	1.03	2.26	0.61	1.61	1.97	1.14
Rumania	11.1	6.6	7.6	2.0	1.2	12.5	9.0	9.6	5.8	4.8	2.92	1.47	0.36	3.28	2.00	1.05
Total six	49.2	48.7	45.9	30.3	28.1	63.1	54.3	49.6	41.4	41.9	1.71	1.66	1.06	2.19	1.82	1.45
Soviet Union	50.8	51.3	54.1	69.7	71.9	36.9	45.7	50.4	58.6	58.6	0.71	0.72	0.97	0.52	0.65	0.82
Total seven	100.0	100.0	100.0	100.0	100.0	100.0	100.0	100.0	100.0	100.0	1.00	1.00	1.00	1.00	1.00	1.00

Source: Trade data: OECE East-West technology transfer data base (Western statistics). NMP weights are those used by the Economic Commission for Europe, quoted from H. Wienert, J. Slater, East-West technology transfer ... op.cit., p. 226 and own computations

From this perspective, the ranking of individual countries is markedly different. The highest technology import intensity for capital goods, throughout the entire period, was registered in Bulgaria, followed by Hungary. These were the only two countries which systematically scored higher than the CMEA average (without the Soviet Union). Rumania reached its peak at the beginning of the 1970s and Poland in the mid1970s. Thereafter, the two countries decreased their technology import intensity dramatically to well below the CMEA average (again exluding the USSR). Czechoslovakia systematically decreased its technology import intensity throughout the entire period, although its point of departure was high, while the Soviet Union behaved quite contrarily, starting form a low level, it systematically improved its position.

As far as the imports of technology-related intermediate inputs are concerned, the highest technology import intensity was recorded by Hungary, followed by Bulgaria. The developments in individual countries largely followed patterns observed in the case of capital goods imports.

An overwhelming part of the technology imported via capital goods was of a material nature (over 90%) and within this category aroung 60% was accounted for by the industry alone (3). Within the industrial sector, producer industries clearly dominated the scene, accounting for over 50% of the total, followed, as a rule, by consumer industries and finally by fuel and energy (see Table 5.3.).

It is interesting to note that while the share of producer industries was fairly constant over the entire period in question (with the exception of Czechoslovakia), that of fuel and energy rose at the expense of the consumer industries. The largest share of producer industries has been regularly recorded in Bulgaria, the Soviet Union and Rumania. Consumer industries, on the other hand, were well represented in the GDR and Hungary. In the fuel and energy industries Czechoslovakia was the leading country.

Comparing the import structure of capital goods from the West with that of the East, we can see that there is a marked assymmetry between the two. Capital goods coming from the CMEA group were predominantly used for fuel and producer industries, whereas those of Western origin were used mainly for producer and consumer industries (Fig. 5.1.).

As fas as the level of sophistication of imported technology is concerned, the available data suggest that the bulk of OECD technology exports to CMEA contained low R+D intensive products

Table 5.3. Eastern Country Capital Goods Imports Identified by Region of Origin, by Broad Industrial End-use Divisions

1. per cent of total industrial capital goods imports from the West

	Fuel and Energy[a]				Producer Industries[b]				Consumer Industries[c]			
	1975	1980	1982	1983	1975	1980	1982	1983	1975	1980	1982	1983
Bulgaria	7	13	13	11	50	68	70	67	43	19	18	22
Czechoslovakia	21	58	27	24	43	20	26	32	36	22	47	43
German Democratic Republic	23	19	13	10	23	46	57	71	54	34	30	20
Hungary	21	19	16	23	39	45	47	41	40	36	28	36
Poland	14	30	36	30	56	55	52	46	31	15	12	23
Soviet Union	15	22	28	37	65	63	56	50	19	16	17	12

2. per cent of total industrial capital goods imports from the East

	Fuel and Energy[a]				Producer Industries[b]				Consumer Industries[c]			
	1975	1980	1982	1983	1975	1980	1982	1983	1975	1980	1982	1983
Bulgaria	38	37	34	39	44	49	51	48	18	14	14	13
Czechoslovakia	35	33	49	48	41	44	33	35	24	24	18	18
German Democratic Republic	35	42	51	50	46	38	31	34	19	21	18	16
Hungary	33	56	59	63	40	28	29	25	27	16	11	11
Poland	27	32	37	35	51	47	36	38	22	21	28	27
Rumania	30	24	25	30	52	52	57	53	18	23	18	17
Soviet Union	28	33	31	33	45	41	41	39	27	26	28	29

a) Fuel and energy branches
b) Metallurgy, engineering, chemicals and construction materials branches
c) Wood, paper, light and food branches

Source: OECD East-West technology transfer data base (Eastern statistics), quoted from: H. Wienert, J. Slater - East-West technology transfer ... op.cit., pp. 264-265

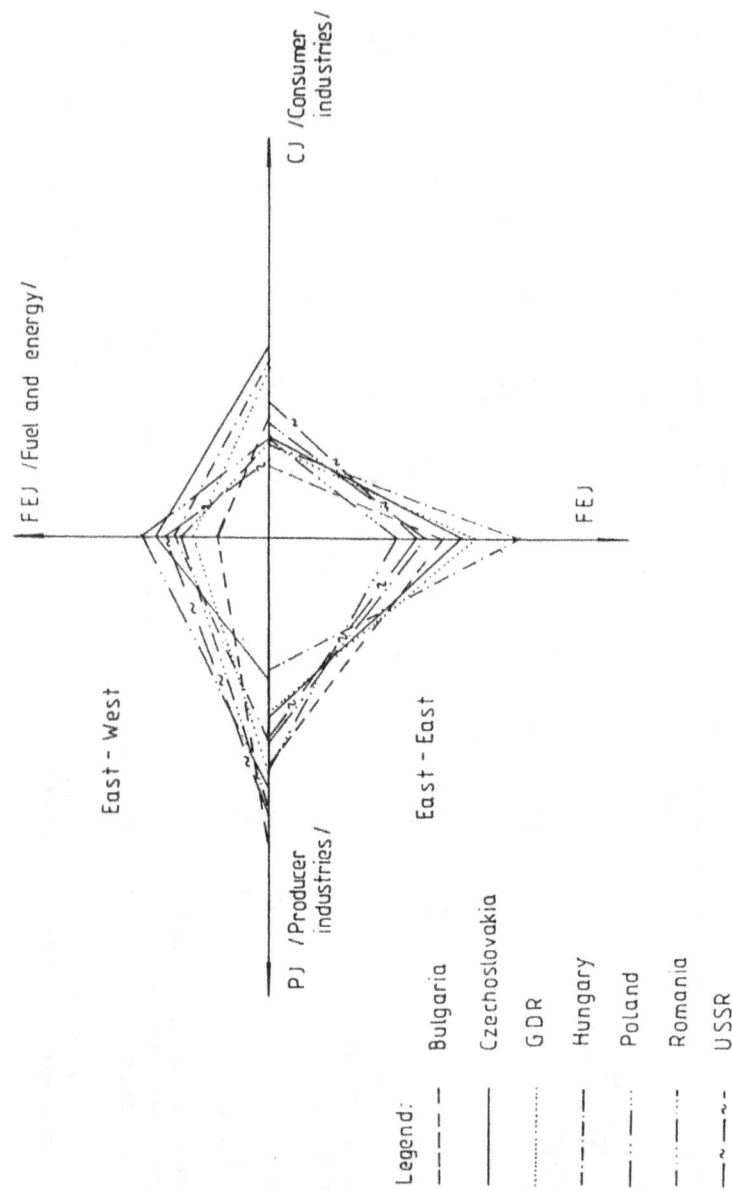

Fig. 5.1. CMEA Import of Capital Goods in the 1970s by End-use Divisions

Table 5.4. OECD Exports of Technology and Technology, Based Products to the Eastern Countries (classified by R+D intensity) (percentage of total embodied technology exports to the East)

	1970	1975	1980	1982	1983
Highly R+D-Intensive					
Bulgaria	2.30	2.89	3.72	4.34	4.64
Czechoslovakia	3.63	3.37	3.35	4.08	4.21
GDR	1.94	1.36	3.25	3.93	2.13
Hungary	3.88	3.50	4.53	4.22	4.22
Poland	2.48	1.82	1.83	2.71	3.26
Rumania	4.12	4.31	6.20	8.33	10.70
Total, excluding the Soviet Union	3.20	2.69	3.65	4.32	4.32
Soviet Union	1.97	1.31	1.53	1.47	1.80
Total, including the Soviet Union	2.67	2.05	2.58	2.56	2.73
Moderately R+D-Intensive					
Bulgaria	20.77	14.39	20.29	19.33	19.87
Czechoslovakia	20.66	25.04	24.14	24.99	26.75
GDR	13.68	20.14	23.41	19.78	18.62
Hungary	28.52	27.83	24.86	25.08	24.36
Poland	21.73	15.20	22.56	23.38	29.03
Rumania	16.14	15.66	18.00	21.05	24.60
Total, excluding the Soviet Union	20.39	18.59	22.29	24.61	24.29
Soviet Union	16.11	13.70	19.15	16.19	17.92
Total, including the Soviet Union	18.55	16.32	20.70	19.40	20.25
Low R+D-Intensive					
Bulgaria	76.93	82.72	75.99	76.34	75.49
Czechoslovakia	75.71	71.60	72.51	70.93	69.03
GDR	84.37	78.50	73.33	76.29	79.24
Hungary	67.61	68.67	70.61	70.69	71.42
Poland	75.79	82.98	75.61	64.91	67.61
Total, excluding the Soviet Union	76.40	78.72	74.06	71.06	71.49
Soviet Union	81.91	84.99	79.32	82.34	80.28
Total, including the Soviet Union	78.78	81.63	76.73	78.04	77.02

Source: OECD East-West Technology Transfer Data Base (Western statistics). Quoted form: H. Wienert, J. Slater - East-West technology transfer ..., op.cit., p. 260

(70-80%), with the share of highly R+D intensive products oscillating around 3-4% of the total (see Table 5.4.).

Interestingly enough, the largest share of highly R+D intensive technology from the total high tech imports was recorded in Rumania throughout the entire period, followed by Hungary until 1980, and thereafter by Bulgaria. The lowest shares (apart form the Soviet Union) were reported in Poland and the GDR. In moderately R+D intensive products, Hungary was at the top, followed by Czechoslovakia until 1982 and by the GDR until 1980.

The Soviet Union confined itself largely to low R+D intensive technologies, which accounted for over 80% of its total imports from OECD.

This peculiar structure of technology import to CMEA could mean that most of its imports have been motivated not by the lack of the relevant know-how, but rather by the lack of the relevant industrial capacities. Consequently, the bottleneck was investments and not the R+D facilities. Such a peculiar structure could also bear the impact of the political and strategic considerations, both of the suppliers and of the recipients.

Hence, one could assume, that the said structure is determined by production capacities on the one hand, and, political strategic considerations on the other.

As far as the sources of the technology inflows ar concerned, the EEC countries have come to play the dominant role, accounting for over 60% of the total, throughout the entire period. Disproportionally large share of both capital goods and technology-related intermediate inputs came from Northern European countries. This can largely be attributed to the political factors, as most of them pursue the policy of political neutrality and remain outside the NATO and Cocom structures. The role of the United States, on the other hand, was extremely minor, particularly in view of their position on the world technology market, which again is a reflection of the political interest pursued by this country (see Table 5.5.).

Japan, however, rapidly increased its share as a technology supplier to CMEA, especially in the case of the Soviet Union, at the expense of the EEC and the United States.

Our picture of the role of pure foreign trade transactions in East-West technology transfer would be incomplete without considering the trade in licenses. It is deliberately discussed at the end of the paragraph as a good part of this trade (between East and West, a rough guess would be around 50%) is already within reg-

Table 5.5. Share of OECD Country Groupings in OECD Exports of Embodied Technology to the East (percent)

	Capital Goods				Intermediate (Technology) Goods			
	1970	1975	1982	1983	1970	1975	1982	1983
Eastern Europe								
EEC 9	73.0	68.9	62.2	58.5	67.2	69.9	68.0	67.6
N. Europe (a)	22.8	20.2	25.1	23.8	22.4	20.7	22.8	22.6
Japan	1.8	5.7	8.3	13.9	5.7	5.3	4.2	5.2
N. America	2.1	4.3	3.4	3.0	2.0	1.7	2.5	2.3
S. Europe (b)	0.2	0.8	0.9	0.7	2.8	2.3	2.5	2.5
Oceania (c)	-	-	-	-	-	-	-	-
Total	100.0	100.0	100.0	100.0	100.0	100.0	100.0	100.0
Soviet Union								
EEC 9	62.5	57.6	38.6	50.7	55.5	62.7	52.9	56.5
N. Europe (a)	23.6	16.4	26.0	27.7	23.8	16.3	19.1	17.3
Japan	10.0	14.7	32.2	19.0	15.8	17.7	19.1	17.3
N. America	3.9	11.2	2.6	1.9	4.9	2.5	6.6	5.8
S. Europe (b)	-	0.1	0.5	0.9	-	0.8	2.3	3.1
Oceania (c)	-	-	-	-	-	0.1	-	-
Total	100.0	100.0	100.0	100.0	100.0	100.0	100.0	100.0

Notes: a) Iceland, Norway, Switzerland, Finland, Austria and Sweden
b) Greece, Portugal, Spain and Turkey
c) Australia and New Zealand

Source: OECD East-West technology transfer date base (Western statistics), quoted from H. Wienert, J. Slater - East-West technology transfer ... op.cit., p. 229

Table 5.6. CMEA Import of Licenses from the West

Year	Czechoslovakia No.	Payments (million $ US)	Hungary No.	Payments (million $ US)	GDR No.	Payments (million $ US)	USSR No.	Payments (million $ US)	Poland No.	Payments (million $ US)
1970	44	46.7	18	1.5	.	.			30	.
1971	32	40.7	29	6.2	36	8.33			34	66.7
1972	44	53.1	35	5.8	18	1.48			57	58.2
1974	34	55.2	43	10.4	14	4.88			70	133.3
1975	46	45.3	60	15.4	13	4.27	700		68	163.7
1976	36	65.8	59	10.0	11	8.42			58	177.6
1977	56	65.0	68	13.9	16	5.70			30	19.1
1978	63	66.8	111	13.6	.	.			30	33.6
1979	57	57.3	96	21.9	.	.			12	49.3
1980	58	46.3	.	22.7	.	.		64.0	6	5.9
1981	42	27.5	.	21.0	.	.			0	.
1982	.	.	.	26.9	.	.			0	.
1983	.	.	.	31.1	.	.			2	.
1970–1983	519	617.5	558	203.7	121	34.9			450	771.8

Source: F. Levcik, J. Skolka – East–West technology transfer. Study of Czechoslovakia, OECD, Paris 1984, p. 37; P. Marer – East–West technology transfer. Study of Hungary 1968-1984, OECD, Paris 1986, p. 119; J. Monkiewicz, J. Maciejewicz – Technology export from the socialist countries, Westview Special Studies in International Economics and Business, Westview Press, Boulder and London, 1986, pp. 64-65 and p. 82

ulated foreign trade transactions, forming the base of subsequent cooperation schemes (see next paragraph).

The statistical base for analyzing the trade in licenses, notwithstanding recent advances, remains much more limited than in the case of the trade in capital goods. Only Czechoslovakia, Hungary and, since recently, Poland publish official statistics which, however, are confined only to some general figures. The other CMEA countries, including the USSR, do not supply any official information on this issue. This, of course, greatly influences the accuracy of the statistics used, however it does not undermine its value and should not cause us to avoid the question at hand.

According to the available data, the number of licenses imported form OECD since 1970 can be estimated at around 2300-2500 of which over 700 were imported by the Soviet Union alone (see Table 5.6.). It was followed by Hungary, Czechoslovakia and Poland. The direct payments made over the period can be estimated at $ US 2.5-3 billion.

Judging form Polish, Czech and Hungarian figures, the bulk of the licenses can be attributed to the machine building (50-60%) and chemical industries (20-30%). Just as in the case of capital goods, major supplying countries are from Western Europe, with the United States playing clearly a secondary role.

The position of the individual CMEA countries in the import of licenses has undergone considerable changes during the period in question. Czechoslovakia seemed to pursue the most stable policy in this area, keeping its imports at a level of 50 to 60 licenses per annums since the mid 1970s. The same is apparently true for the Soviet Union, although no precise information is available. Poland expanded its import of licenses by the mid 1970s, to decrease it dramatically thereafter. Hungary, on the other hand, is still expanding its imports, although they are already relatively high.

3. Regulated Foreign Trade Transactions. The Case of Interfirm Cooperation

East-West industrial cooperation agreements (EWIC) developed rapidly in the first half of the 1970s, parallel to the expansion of the technology import programs of the CMEA countries. At the end of 1975, the number of agreements was estimated at ca. 1000 and apparently it has remained at the same level since then, with a slight decline in 1981-1984. Thus, it is clear, that the agreements represent a margin of the total number of commercial transactions. Their limited importance is well illustrated by the fact that the trade flows they animated were not more than 5-7% of the East-West total (4). As they are so insignificant in quantitative terms, it is believed that EWIC's will open new qualitative possibilities for transborder technology flows between the West and East. This is attributed basically to the long-term nature of EWIC's, which presupposes regular learning processes, and to the existence of stronger incentives for the Western partners for the successful transmission of their know-how. The peculiar nature of the financial arrangements of EWIC contracts, based foremostly on some buyback elements, is an additional factor promoting technology flows, due to the increase in hard currency earnings of the CMEA countries.

Table 5.7. Industrial East-West Cooperation Contracts (EWIC) According to Institutional Forms (% of the total)

	1972	1979	1982	1984
Licensing	28.2	15.0	13.6	13.1
Turn-key deliveries	11.9	20.0	19.6	16.5
Specialization	4.0	4.1	3.3	3.7
Co-production	26.2	22.6	27.7	27.0
Co-production and specialization involving R+D only	6.9	12.4	10.3	12.9
Sub-contracting	7.9	3.3	10.3	12.9
Joint ventures	14.9	13.1	12.7	13.3
Tripartite cooperation	-	9.5	10.3	10.4

Source: J. Stankovsky - Gemischte Unternehmen als Element wirtschaftlicher Zusammenarbeit zwischen Ost und West, Oktober 25, 1983, mimeo, p. 37 and own computations

The most popular forms of EWIC are co-production and specialization arrangements, amounting to ca. 40% of all EWIC contracts throughout the 1970s. They are based on the reciprocal
supply of goods and services necessary for the final product (coproduction) or on completing each partners' range of products
(specialization). This can comprise manufacturing activities as well
as research and development undertakings.

The next most widespread form of EWIC are turnkey deliveries
in exchange for products or components, which accounted in 1979
for 20% of the total, declining thereafter to 16.5% in 1984 (see
Table 5.7.).

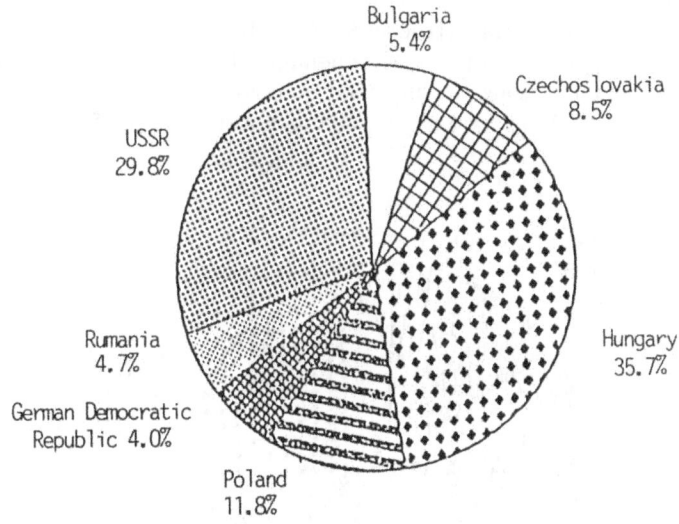

Fig. 5.2.   Breakdown of Industrial Cooperation Contracts by
            Country, 1984 (percentages)

The total value of the Western supplies under this form of cooperation from 1970-1980, is estimated at around $ 30 billion,
whereby equipment worth over $ 19 billion was scheduled for the
Soviet Union alone, followed by Poland (over $ 5 billion) and the
GDR ($ 3.6 billion) (5).

The third mostly practiced form of EWIC is the supply of licenses, with payment in the resulting products manufactured under the license, followed by joint venture arrangements (equity links) and tripartite cooperation ventures.

From among the CMEA countries, Hungary has made the most acitve use of EWIC's, accounting regularly for ca. 1/3 of the total number of contracts concluded. It was followed by the Soviet Union (below 30% of the total), and subsequently by Poland (see Fig. 5.2.).

The distribution of EWIC contracts among the individual CMEA countries was fairly stable during the 1970s, thereafter, however, there was an increase in Soviet involvement and a decrease in Polish participation. Both changes, however, were not of a dramatic nature.

As far as the industrial structure of EWIC is concerned, three sectors seem to play a major role: chemical industry (over 1/5 of the total), mechanical engineering and machine tools (slightly below 1/5 of the total) and transport equipment (see Fig. 5.3.).

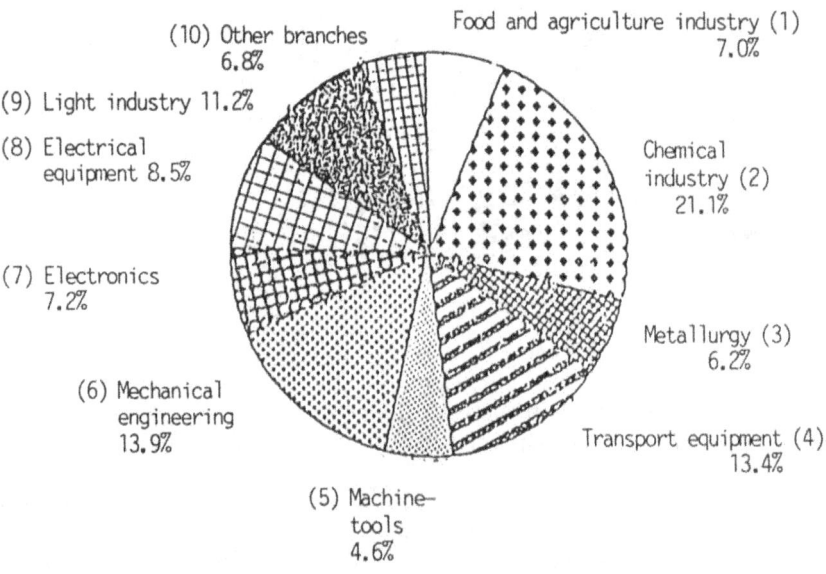

Fig. 5.3.   Breakdown of Industrial Cooperation by Industry
(percentages)

Table 5.8. EWIC by CMEA Country, Type of Contract and Industry as of 1984 (percent)

Country	Grand total	Type of contract							Industry										
		Total	A	B	C	D	E	F	Total	1	2	3	4	5	6	7	8	9	10
Bulgaria	5.4	100.0	4.5	13.6	50.0	4.5	18.2	9.1	100.0	9.1	9.1	4.5	6.8	4.5	25.0	11.4	6.8	20.4	2.3
Hungary	35.7	100.0	20.3	7.5	36.4	6.2	17.2	12.4	100.0	8.9	18.2	3.1	8.6	4.8	11.7	7.6	14.1	15.8	7.2
Poland	11.8	100.0	19.8	25.0	33.3	1.0	5.2	15.6	100.0	4.2	17.7	12.5	15.6	2.1	16.7	12.5	5.2	6.2	5.2
GDR	4.0	100.0	6.1	54.5	12.1	3.0	6.1	18.2	100.0	3.0	15.1	6.1	27.3	9.1	18.2	9.1	-	9.1	3.0
Rumania	4.7	100.0	5.3	7.9	15.8	2.6	57.9	10.5	100.0	2.6	34.2	10.5	3.2	-	-	5.3	2.6	5.3	5.3
Czechoslovakia	8.5	100.0	20.3	1.4	69.6	-	7.2	1.4	100.0	10.1	8.7	-	11.6	16.0	21.7	2.9	14.5	11.6	2.9
USSR	29.8	100.0	3.7	24.7	56.4	0.8	6.6	7.8	100.0	4.9	35.0	9.9	7.8	5.3	13.6	5.8	3.3	5.8	8.6
Total	100.0	100.0	13.1	16.5	43.6	3.1	13.3	10.4	100.0	6.8	22.5	6.5	11.3	5.5	14.4	7.4	8.3	10.8	6.5

Legend:
A – Licensing
B – Turn-key plants
C – Coproduction and specialization
D – Sub-contracting
E – Joint-ventures
F – Tripartite cooperation
1. Food and agriculture
2. Chemical industry
3. Metallurgy
4. Transport equipment
5. Machine tools
6. Mechanical engineering
7. Electronics
8. Electrical equipment
9. Light industry
10. Other

Source: ECE, TRADE/R.487, October 19, 1984, pp. 9-10

The situation, however, is fairly diversified from country to country. Thus, Bulgaria favored foremostly mechanical engineering and light industry, Hungary - chemicals, light industry and electrical equipment; Poland - mechanical engineering, chemical industry and transport equipment; GDR - transport equipment and mechanical engineering; Rumania - chemical industry and transport equipment; Czechoslovakia - mechanical engineering and machine tools and the Soviet Union - chemical industry and mechanical engineering (see Table 5.8.).

Large inter-country differences could be also observed in the structure of EWIC, by type of contracts. Thus, for example, Bulgaria favored primarily coproduction and joint ventures; Hungary, coproduction, licenses and joint ventures; Poland - coproduction, licenses and turn-key plants, GDR - turn-key plants and tripartite cooperation; Rumania - joint ventures and coproduction; Czechoslovakia - coproduction and licenses and the Soviet Union - coproduction and turn-key plant deliveries.

As far as the Western countries are concerned, the most active participants in EWIC's are the European countries, with West Germany accounting for over 1/4 of the total. It is followed by Italy (11.7% in 1984), France (9.7%), U.K. (7.3%) and Austria (7.1%). The United States account for 11.6% of the total, which is clearly above their share registered inpure foreign trade transactions, whereas Japan, with 5.7% of the total, is below the level recorded here (6).

4. Conclusions

The foregoing discussion indicates that there was a significant increase in Western technology infusion into CMEA countries during the 1970s. The entire period was clearly split up into distinct sub-periods: before and after the mid 1970s. Between 1970-1975, the value of technology infused more than doubled, thereafter, however, it decreased in real terms. This decrease was especially dramatic and prolonged in the case of Eastern Europe. The principal reason for this was the aggravated balance of payment problems in most of the CMEA countries. Under these circumstances, the capital goods imports were the most severely hit. On the other hand, the import of technology-related intermediate inputs suffered much later and to a much lesser extent. Interestingly enough, even in the expansionary period 1970-1975, the dynamics of East-West technology flows remained below that observed in the overall Western exports to CMEA.

This indicates that the import of technology was not one of the top priorities of the CMEA policy makers during this period, which is so frequently assumed, and that apparently a much broader range of other social and economic goals has been pursued. In absolute terms, around half of the technology imported from the West went to the Soviet Union, followed by Poland and Czechoslovakia.

In relative terms, however, major beneficiaries of Western technology inflows were Bulgaria and Hungary, with the Soviet Union clearly at the bottom of the ranking list. Contrary to popular expectations, most of the technology infused was of low R+D intensity; the high-tech ratio, on the other hand, was strikingly low, which could mean that the impact of imported technology on the CMEA technological advancement must have been limited, no matter how effectively it was utilized.

An overwhelming share of the imported technology was destined for industry - a traditional priority of CMEA economic policies. Within this sector, producer industries received the lion's share, the rest being split equally between consumer industries and the fuel and energy branches. In the first half of the 1970s, consumer industries clearly ranked second (after producer industries), thereafter, however, due to growing balance of payment problems and the increased deficits in the regional energy supplies, they were largely replaced by the fuel and energy branches.

The foregoing analysis pointed out that there were substantial differences among the individual CMEA countries, with regard to their policies vis-a-vis the technology importation from the West. These differences were, first of all, reflected in the varying intensity of the actual relative technology inflows, if measured by national net material products.

The inter-country differences manifested themselves also in the destination of the technology import. Thus, Bulgaria, the GDR and Hungary gave much more attention to consumer industries (as measured by the capital goods import structure), whilst the Soviet Union and Rumania clearly preferred the producer industries. Poland and Czechoslovakia, on the other hand, were somewhere in the middle.

Notable differences also existed in the level of sophistication of the imported technologies. Judging from the available dat, Rumania had the highest share of most-sophisticated technologies in its technology import. It was followed by Hungary, and, as of recently, by Bulgaria. Hungary and Czechoslovakia, on the other hand, had the largest shares of medium level technologies throughout the 1970s. As far as low level technologies are concerned, the Soviet Union was clearly in the lead, followed closely by Poland and Rumania.

Finally, there were also substantial differences in the institutional forms of technology import, and particularly in the degree of application of the regulated forms of technology transfer. This can be seen clearly when the countries' shares in industrial cooperation contracts are compared with the respective countries' shares in regional net material products (coefficient of relative regulation). As the calculations show, a clear leader in this field was Hungary, with a relative share nearly 14 times higher than the CMEA average. It was followed by Bulgaria - 2.3 above the CMEA average, Czechoslovakia (1.8) and Poland (1.6) (all figures refer to 1980-1982). The lowest propensity in regulated forms was demonstrated by the Soviet Union (0.42 of the CMEA average) and the GDR (0.67). Further differences could also be detected in the structure of the regulated forms in use.

Thus, it seems safe to conclude that the technology import policies of the individual CMEA countries, although showing some degree of similarity were extremely differentiated. The reason for these differences can be partly attributed to the institutional mechanisms prevailing, to the different mix of goals pursued, to the different resource endowment, the different political standing in the East-West relations, etc.

Throughout the entire period in question, the principal methods of technology import from the West were the traditional pure foreign trade transactions and especially the import of capital goods. Regulated forms of technology infusion remained quantitatively insignificant, although they developed rapidly during the period concerned.

Nonetheless, the emergence of regulated forms of technology flows in East-West relations must be interpreted as the increased readiness of the CMEA countries to give up some of their classic ideas of the foreign trade and technological sovereignty, in the anticipation of a better access to Western technological resources and higher learning through cooperation benefits.

This also paved the way for the subsequent emergence of the direct East-West equity link, which will be at the center of discussion in the next chapter.

References and Footnotes

1   H. Wienert, J. Slater - East-West technology transfer. The trade and economic aspects, OECD, Paris, 1986, p. 200

2   H. Wienert, J. Slater - East-West technology transfer ... op.cit., pp. 200-201

3   H. Wienert, J. Slater - East West technology transfer ... op.cit., pp. 212-215

4   UNCTAD TD/B/859, August 7, 1981, p. 29

5   Economic Commission of Europe, TRADE/R. 484, September 10, 1984, p. 15

6   ECE, TRADE/R. 487, October 19, 1984, p. 11

Chapter 6
Foreign Direct Investments and East-West Technology Transfer

1. Introduction

The bulk of today's technological changes in Western economies is carried out by large industrial entities, predominantly of an international nature. These transnational corporations are also - as indicated in Chapter 4 - responsible for most of the international technology transfers which primarily occur through foreign direct investments.

It is frequently argued that the movement of technology across borders is inseparable from the transborder movement of investment capital and thus, obstacles of foreign investments automatically become obstacles of international technology flows. This issue is of particular interest to the socialist countries due to their peculiar position with respect to foreign direct investments. This position has been always precarious and vague. On the one hand, most of the theorists and policy makers in these countries took for granted the fact that economic processes objectively tend towards internationalization, and that no national or international action could stop this phenomenon in the long run. On the other hand, however, the same theorists and policy makers seemed to believe that the process of internationalization could be stopped half-way and, thus, the internationalization of productive capital

---

* This chapter draws heavily on my joint study with A. Wass von Czege - Western direct investments in CMEA and technology transfers: some preliminary observations, IAuU, Universität Hamburg, Forschungsbericht Nr. 44, July 1986

could be avoided or at least substituted by some other measures (for example, industrial cooperation arrangements).

These views were heavily ciriticized by the proponents of the foreign direct investments, who claimed that nothing of this sort is possible and that the sooner foreign capital becomes accepted, the better it is for the socialist countries (1).

In discussing the relationship between foreign ownership and technology transfer, two principal questions should be addressed:

a) What are the reasons for internalizing the technology transfer transactions?
b) What consequences could arise for the technology recipient if internalization is not permitted or restricted by him?

In the following, we shall cover these individually.

2. Available Theoretical Explanations

There are three major theories explaining the internalization of international technology transfer: one associated with the names of Williamson, Buckley and Casson, the other elaborated by S. Magee and the third put forward by R. Findlay (2).

The Williamson-Buckley-Casson explanation is based on the theory of the firm developed by O. Williamson, within the framework of the so-called new institutional economics. According to this theory, the modern company evolved primarily as a result of the attempts to economize on transaction costs. All activities (transactions) which are more effectively arranged within hierarchical structure than via a market mechanism become internalized within the company (3).

Technology transactions represent a typical case for internalization which, in terms of this theory, may be explained by "information impactedness" (i.e. the information advantage of one contracting party over the other), high uncertainty of the transactions (in terms of both the final results and the transaction process itself), its low recurrency (i.e. low repetitiveness or routinization) and a high share of specialized human and capital investments (4).

All of these reasons provide a strong stimulus for a firm to internalize the technology creation as well as its transfer, both at home and internationally. The theory implicitly suggests that the supply of new technology through international markets is significantly limited in comparison to the supply via foreign direct investments, and also that with the increasing standardization of technology (i.e. aging) the incentives for its internalization diminish, and hence arm-length agreements are more likely to occur.

The theory assumes that the situation described above results from the objective characteristics of technology transactions and the technology market. Precisely, this principal contention has been challenged by S. Magee in his theory of appropriability of technology creation (5). According to him, market failures to organize technological transactions result from the conscious manipulation of multinational companies (MNCs) which, already during the process of technology creation, tend to produce things which are difficult to duplicate without their direct involvement. The reason behind this attitutde, according to S. Magee, can be explained by the MNCs desire to receive a maximum return on their research and development investments.

R. Findlay, on the other hand, associates the necessity for internalizing technology transfer basically with the problems of effective communication (6). He argues that in the past it was the migration of individual inventors which played a major role in international technology diffusion. Todays, with R+D being performed primarily by large R+D departments, the role of individuals is taken over by multinational corporations. Apart from the positive communication effects, the technology transfer via foreign direct investments additionally creates demonstration effects within the recipient economy and enforces local competition, thus contributing to local technological learning and innovation efforts. Therefore, the more foreign direct investments (FDI) there are, the more rapid is the recipient's technological catching up (7).

All of these concepts, though proposing different interpretations of the reasons for the internalization of technological transfer transactions, arrrive at similar conclusions. These can be summarized briefly as follows:

1. Technology transfer transactions are more eagerly handled by technology owners via hierarchies (through FDI) than via markets.

2. This is especially true in the case of new technologies and is differentiated according to the nature of the technology. It is argued that due to the internal characteristics of technology,

the inclination to internalize is more profound in the case of technological products than processes (8).

3. Technology transfer proceeds more smoothly and brings more technological benefits for the receiving economy when internalized.

Based on this, one can argue that as far as technology recipients are concerned, the following hypothesis could be formulated:

1. Removing or limiting foreign ownership decreases the volume of foreign technologies available (9).
2. Removing or limiting foreign ownership provides access largely to outdated and simple technologies.
3. Removing or limiting foreign ownership results in a concentration of foreign technology inflow on process-centered technologies.
4. Removing or limiting foreign ownership results in a lower effectiveness of technology transfer and limits its effects.

Before going into the further analysis of the CMEA case, let us briefly summarize the results of the empirical findings in this area.

3. Foreign Direct Investments and Technology Transfer: Empirical Findings

Currently, empirical observations are very scarce indeed. In fact, they are limited to the studies by E. Mansfield et al., E.K.Y. Chen, C.C. Coughlin and S. Globerman. The first one dealt with technology owners, the others with the experience of technology recipients.

The study by E. Mansfield, A. Romeo and S. Wagner was based on the interviews of 30 US industrial companies which were asked, inter alia, about their anticipated actions with respect to the utilization abroad of their R+D projects currently under way (10). The companies surveyed included 20 industrial firms active in fabricated metal products, machinery, instruments, chemicals, paper and the tire industries, and 10 major chemical firms. The results of the survey are reproduced in Table 6.1.

Table 6.1. Distribution in percent of R+D Projects, by Anticipated Channel of International Technology Transfer, First Five Years after Commercialization, 23 Firms, 1974 (a)

	Foreign subsidiary	Exports	Licensing	J-Vs	Total (b)
All R+D projects					
16 industrial firms	85	9	5	0	100
7 major chem. firms	62	21	12	5	100
Projects aimed at (c)					
entirely new products	72	4	24	0	100
product improvement	69	9	23	0	100
entirely new process	17	83	0	0	100
process improvement	45	53	2	1	100
Projects of estimated rate of return (d)					
less than 20%	36	19	38	7	100
20-39%	46	29	19	5	100
40% and more	100	0	0	0	100

(a) only projects where foreign returns are significant (more than 10% of the total sales for 16 industrial firms and 25% of the total sales for chemical firms)
(b) rounding
(c) only 6 chemical firms could be included
(d) only 4 chemical firms could be included

Source: E. Mansfield, A. Romeo, S. Wagner - Foreign trade and US research and development, in: The Review of Economics and Statistics, Vol. 61, 1979, p. 55

The findings generally confirm the hypotheses no. 1. to 3., however, with one important qualification. It seems that there is a discontinuity in the appreciation of foreign ownership by US companies: they want either all or nothing. Surprisingly enough, the companies generally preferred straightforward licensing to joint-venture arrangements. Thus, it seems that allowing foreing ownership is not enough, the degree of freedom in this respect is also important.

Table 6.2. Type of Technology Transferred

| Country | % of total | | No of |
	Product	Process	Observations
Yugoslavia/Spain	27.5	72.5	120
Greece/Portugal	45.5	54.5	22
France	22.1	77.9	263
FRG	28.6	71.4	234

Source: C.C. Coughlin, The relationship between foreign ownership and technology transfer, Journal of Comparative Economics, 7, 1983, pp. 400

The study by C. Coughlin, based on data collected by W. Davidson and R. Vernon, was concerned with the effect of national regulations of foreign direct investments on technology transfer. It covered 57 US multinational corporations and the overseas spread of their 406 innovations and 548 imitations between 1945-1975. Thus, contrary to E. Mansfield et al., C. Coughlin investigated past performance on a country wise basis. The countries covered included Yugoslavia and Spain, as representatives of a more restrained approach towards FDI, and Greece and Portugal as representatives of a liberal approach. It also covered France (hard line on FDI) and West Germany (liberal approach). The results of the calculations are reproduced in Tables 6.2. and 6.3.

The figures appear to indicate that with the growing restrictiveness of national legislation on foreign investments, the technology transferred is disproportionally more process- than product-oriented (hypothesis no. 3) and that it becomes older (hypothesis no. 4).

Table 6.3. Age of Technology Transferred (a)

Country	Type of Tech- nology	Mean of Speed/ Years	Standard Devia- tion	No. of Obser- vation
Yugoslavia/Spain	Product	14.1	14.6	33
	Process	15.8	11.0	87
Greece/Portugal	Product	10.2	5.3	10
	Process	15.1	16.2	12
France	Product	10.0	8.3	58
	Process	10.1	8.7	205
FRG	Product	8.9	9.9	67
	Process	11.4	9.6	167

(a) Measured as the difference between the year a specific tech- nology was transferred and the year it was first commercial- ized by the firm

Source: C.C. Coughlin, The relationship ..., p. 412

Unlike E. Mansfield and C.C. Coughlin, S. Globerman and E.H.Y. Chen were interested in the speed of diffusion of foreign technology, depending on the degree of foreign capital penetration. Investigating this question in the Canadian tool and die industry, S. Globerman did not detect any positive correlation in this re- spect (11).

E.K.Y. Chen, on the other hand, demonstrated a positive cor- relation between the degree of foreign capital penetration and the rate of technology diffusion in the Hong Kong manufacturing in- dustry (12).

With all these theoretical hypotheses and findings in mind, let us now try to review the experiences gathered so far by CMEA countries in this area.

4. Foreign Direct Investments (FDI) and Technology Transfer to CMEA: Trends and Patterns

CMEA countries act as both home and host countries for foreign direct investments. Investments made by CMEA countries have a relatively long history, however, their dynamic expansion only took place in the 1970s. This growth can be explained basically by the attempts undertaken by several of the CMEA countries to build-up their exports and to rationalize their production structures. One of the instruments employed were equity ventures abroad as an institutional innovation in the foreign economic activities of socialist countries.

There are no official statistics on the foreign investments made by CMEA countries. However, due to the pioneering work of McMillan, the overall picture is more or less known. According to him, at the end of 1983, there were around 650 enterprises with CMEA direct capital involvement world-wide, of which ca. 440 were in the West and 213 in the South. The total value of the capital invested in the West was around $ 0.5 billion (13). Between 1977 and 1983 the number of companies in Western countries with East European participation increased by 50%. This was principally due to the performance of the GDR, Czechoslovakia and Hungary. The major investors were the USSR (111 cases in 1983), followed by Hungary (107) and Poland (102). As far as the host countries were concerned, West Germany dominated with 83 companies. It was followed subsequently by the United Kingdom (68 companies) and Austria (44 companies) (see Table 6.4.).

Most of the companies established in Western countries are purely trading ventures. Less than 10% are engaged in material production. Thus, the possibility of direct contribution to eventual technology transfer to the home countries, is negligible. Indirectly, however, these companies may play a certain role both as the source of information on new technologies available and as the trading agent.

As far as Western investments in CMEA are concerned, they are of a more recent nature, the first having taken place at the beginning of the 1970s. Rumania received the first (1973), followed subsequently by Hungary (1974), Poland (1976), Bulgaria (1980), Czechoslovakia (1986) and finally the USSR (1987).

The principal policy of all CMEA countries so far, has been a limitation of the share of foreign capital up to 49% of the set up ventures. The only exclusion is the 1982 Polish law on foreign

Table 6.4. Companies with East European Partners by Host and Home Country, 1977 (a), 1981, 1983

Host Country	Bulgaria 1977	81	83	Czechoslovakia 77	81	83	GDR 77	81	83	Poland 77	81	83	Hungary 77	81	83	Rumania 77	81	83	USSR 77	81	83	Total 77	81	83
Australia	0	0	0	0	1	1	0	0	0	0	3	4	0	0	0	0	0	0	0	4	4	0	8	9
Austria	2	2	2	0	1	1	3	3	3	4	8	8	13	16	23	2	1	3	3	5	4	27	34	44
Belgium	1	2	2	2	2	3	1	2	7	6	7	8	0	1	1	0	0	0	10	10	13	20	24	34
Denmark	0	0	0	0	0	0	0	1	1	1	1	1	1	1	2	0	0	0	0	2	2	2	5	6
Finland	0	0	0	0	0	0	0	1	1	1	1	1	1	1	2	6	5	7	7	8	9	9	11	13
France	3	5	4	5	3	5	4	3	3	7	8	8	5	3	7	0	1	2	12	12	12	42	39	46
Greece	0	2	2	0	0	0	0	0	0	0	0	1	0	1	1	0	0	0	0	2	2	0	8	8
Spain	1	1	1	0	1	0	0	0	0	2	2	2	1	1	3	1	1	1	3	6	6	8	13	13
Netherlands	1	1	1	1	1	1	1	2	2	3	3	3	3	3	3	0	0	0	4	2	2	14	12	12
Ireland	0	1	1	0	0	0	0	0	0	0	0	0	0	1	1	0	0	0	0	0	0	0	2	2
Japan	2	2	3	0	1	0	0	0	0	1	0	1	1	1	1	0	0	0	1	1	1	5	6	7
Lichtenstein	0	0	0	0	0	0	0	0	0	0	0	0	0	1	1	0	0	0	0	0	0	0	1	1
Luxembourg	0	0	0	0	0	0	0	0	0	0	0	0	1	1	1	0	0	0	0	1	1	2	2	2
Canada	1	1	2	2	5	5	0	0	0	2	3	4	1	1	1	2	1	2	5	5	5	13	16	19
New Zealand	0	0	0	0	0	0	0	0	0	0	0	0	0	0	0	0	0	0	0	3	3	0	4	4
Norway	0	2	2	0	0	0	0	0	0	2	1	1	0	0	0	0	0	0	3	3	3	5	6	7
Portugal	0	0	0	0	0	0	0	0	1	0	1	1	0	0	0	0	0	0	0	0	0	0	1	1
GFR	11	12	13	1	1	6	0	0	1	12	16	17	13	15	28	7	7	7	10	12	11	54	63	83
United St.	0	0	0	0	0	2	0	1	1	7	13	16	2	7	10	2	2	1	5	5	5	16	32	35
Sweden	1	0	0	3	3	3	1	1	2	5	5	7	3	2	3	0	0	0	4	4	5	17	15	21
Switzerland	2	2	2	0	0	2	0	0	0	2	3	3	2	2	2	1	2	2	1	4	4	8	14	15
U.K.	4	6	6	10b	9	10	5b	8	9	8	15	13	8	5	10	3	5	5	10	14	15	48	59	68
Italy	6	5	6	2	2	3	0	1	1	2	3	3	3	3	6	5	6	5	6	8	8	24	28	31
Total	35	44	48	26	32	44	15	23	31	65	96	102	58	68	107	29	32	36	84	111	116	313	403	484b

a) including banks and financial institutions
b) of which 48 are no longer operational

Source: Transnational Corporations in World Development. A Re-Examination. UN CTC, New York 1978, p. 283; C.H. McMillan – Growth of External investments by the Comecon countries, The World Economy, Vol. 2, No. 3 (September 1979), p. 365; C.H. McMillan – The Foreign Investment Activity of the Comecon Countries: Actors and Strategies. University of Reading Discussion Papers in International Investments and Business Studies, No. 73, June 1983, p. 1a; C.H. McMillan – Multinationals from the second world, MacMillan Press, London 1987, p. 34

investments (on small scale ventures), which envisages, as a rule, wholly-owned undertakings. In two other countries, Hungary and Bulgaria, it is envisaged that in special cases, the famous 49-51% rule can be altered.

The inflow of Western capital so far is moderate if not negligible. In Rumania, up to 1987, nine joint-ventures were set up, of which four were subsequently dissolved. On the whole, the value of foreign investments (both in cash and in kind) may be estimated at around $ 40 million (14) (Table 6.5.).

Hungary was much more successful than Rumania. Up to 1986, a total of 60 joint-ventures were set up and the cumulated book value of foreign investments is currently estimated at $ 80 million (15). Half of the investors come from West Germany (15) and Austria (14), followed by Switzerland (8) and the USA (6) (see Table 6.6.).

Up until 1980, the dynamics of foreign investments in Hungary was limited. Due to a gradual modification of the relevant legislation, as well as to the more aggressive marketing strategy of the Hungarian authorities, the number of companies with Western participation increased 7.5 times in 1981-86. Particularly large increases occurred after 1984, when far-reaching modifications in the existing law on foreign investments were introduced.

In Poland, there are currently 695 companies with Western participation, most of them wholly-owned by foreign investors. Their overall employment reached the level of ca. 57.000 persons with an output of ca. $ 516 million (16). The total value of foreign capital invested is approximately $ 50-100 million (17).

Up to 1986, Polish legislation allowed only for small scale investments. Since June 1986, with the enactment of a new law on foreign investment, other types of foreign investments have also become acceptable. Up to the end of 1987, seven joint-ventures between the state enterprises and foreign investors were approved, with a total value of anticipated foreign investment of about $ 20 million.

Thus, of all the CMEA countries, Poland seems to have the largest experience in handling foreign investments and, at the same time, is the most heavily affected by them.

Bulgarian's experience as a host country for foreign direct investments is limited to 9 ventures (see Table 6.7.).

Table 6.5. Current List of Capital Venture Projects Located in Rumania

Joint Company	Year of Estab-lishment	Western Partner and Share	Rumanian partner(s) and Share(s)	Capitali-zation	Sector
1. ROM CONTROL DATA	1973	Control Data Corporation (USA) – 45%	Industrial Center for Electronics Technology and Com-puters – 55%	US $ 4.0 million	Manufacture and marketing of peri-pheral equipment for computers
2. RESITA-RENK	1973	Benk A.G. (FRG) – 49%	Resita Engineering Works – 40% Uzinexportimport – 11%	DM 20 million	Manufacture of gears and marine transmissions
3. RIFIL	1973	Romalfa (Italy) – 48%	Industrial Center for Synthetic Fibres – 52%	US $ 2.3 million	Manufacture of acrylic fibres
4. OLTCIT	1977	Citroen (France) – 36%	Industrial Center for Passenger Cars and Com-Commercial Vehicles – 49%	F 500 million	Manufacture of passenger cars and replacement parts
5. ROLISHIP		National Com-pany Maritime Transport (Libya) – 49%	Navrom – 51%	US $ 2.1 million	Maritime shipping

Source: ECE TRADE/R.489 p. 31, ECE TRADE/AC.21/R.1 pp 7-9; various press reports

Table 6.6. Joint-ventures in Hungary

Investing Country	No. of Companies with Western participation	up to 1980	1981-83	1984-86
West Germany	16	2	4	10
Austria	14	2	3	9
Switzerland	8	3	2	3
USA	6	1	1	4
Sweden	3	1	1	1
France	2	-	1	1
U.K.	2	-	-	2
Denmark	2	-	-	2
Finland	2	-	-	2
Saudi Arabia	1	-	1	-
Belgium	1	-	1	-
Greece	1	-	1	-
Japan	2	1	-	1
Italy	1	1	-	-
Netherlands	1	-	-	1
Portugal	1	-	-	1
India	1	-	-	1
USSR	1	-	-	1
Total	60 (a)	8 (a)	15 (a)	37 (a)

(a) The columns do not add up, as in some ventures more than one country has participated

Source: Data supplied by Hungarian Chamber of Foreign Trade

Table 6.7. Western Investments in Bulgaria

Joint Company	Year of Establishment	Western Partner(s) and Share(s)	Bulgarian Partner(s) and Share(s)	Capitalization	Sector
1. Fanuc Machinex	1981	Fujitsu Fanuc (Japan) - 50%	Machinoexport FTO ZMM (manufacturing Kombinat) - 50%; IZOT (manufacturing Kombinat)	leva 700.000	Service and maintenance of automation controls and technical consultation services
2. Tangra	1981	Tangra S.A. (Switzerland) - 20%	Neftochim Burgas Cnimimport FTO - 80%	n.a.	Manufacture and export of plastic and metal products, including ballpoint pens
3. Sofia Mitsukoshi	1982	Maruichi Sjoha (trading house) Mitsubishi department store chain - both to Japan to Japan together - 49%	11 enterprises, including: 4 FTOs (Industrialimport Vinimpex, Intercommerce and Corecom) 2 retail chains 5 manufacturers Total share: 51%	n.a.	Design, development, manufacture, distribution and export of consumer goods; also operates as an agent with foreign trading rights
4. Systematics	1984	Honeywell - 40% (USA)	Ministry of Chemical Industry - 40% Chimimport - 10% Chimkomplet - 10%	n.a.	Marketing, technical assistance and training in automation and control
5. Chimtrade	1984	Dow Chemical (USA)	Chimimport Chimkomplet	n.a.	Marketing and trade in chemical products
6. Futex	1984	Fukazawa Chemical Laboratory (Japan)	Teknika	n.a.	Production and trade of oil and fuel additives
7. ESE/Elprom-Somel Electroimpex	1984	Somel (France)	SEC Elprom Electroimpex	n.a.	Planning, technical assistance, training, trade in the field of electrotechnics and machine tools
8. APV-Bioinvest	1984	APV (U.K.) - 51%	Bioinvest - 49%	200.000	Marketing and trade in pharmaceutical products, food processing confectionary dairy and brewing sectors
9. Tobumex	1985	Japan	n.a.	n.a.	Engineering services in metallurgy

Source: Ostwirtschaftsreport, various years; TRADE/AC.21/R1 p. 3. Presseschau Ostwirtschaft, various years

Four ventures were set up with Japanese companies, two with US firms and the remaining with the participation of French, Swiss and U.K. companies.

So far, Czechoslovakia has only one joint-venture with Western participation (set up in 1986) and the Soviet Union has four, all established in the first half of 1987.

Let us turn our attention to the type of technology being transferred via foreign investments to the CMEA countries. Here, two approaches are possible. We can either analyze the technology intensity of the foreign investments per se, or try to determine what their relative technology intensity is in comparison with other forms of technology inflows to the CMEA region.

The technology intensity of foreign investments can be assessed on the basis of their sectoral distribution. It is assumed that the more projects there are in high-tech industries, the higher their technology intensity (see Table 6.8.).

As indicated by figures provided in Table 6.8., most of the foreign investments so far are in services, followed by textiles and leather industries, and only thereafter by chemicals and pharmaceuticals. High-tech industries such as electronics/EDP, electrical equipment or machinery and transport equipment do not come to the forefront. Thus, it seems that the technological intensity of the said investments is low. This contention is well supported by our previous observations, indicating that the existing joint-ventures are, by and large, relatively small units. This, in turn, presupposes that they are based primarily on small-scale, non-capital intensive technologies. With this type of investments only secondary technologies are available.

In assessing the relative progressiveness of the technologies infused through foreign investments, an index of branch comparative advantages of foreign investments in relation to total East-West industrial cooperation arrangements (EWIC), has been calculated pointing out the branches in which foreign investments have been in use more or less than other forms of EWiC.

The picture which emerges is quite astonishing in view of the hypothesis spelled out earlier (Table 6.9.). In general terms, it seems that the technologies supplied via FDI were more outdated than those via arm-length arrangements. On the one hand, FDI demonstrated their comparative advantages for the CMEA countries in food and agriculture, light industry and construction and services. On the other hand, FDI's were clearly under-represented

Table 6.8. Distribution of Joint Ventures in Eastern Europe by Economic and Industrial Sectors (a)

Sector	Bulgaria (1986)	Hungary (1986)	Rumania (1986)	Poland (1985)
Food and agriculture		3		61
Chemicals and pharmaceuticals	1	5	2	215
Extractive industries including metallurgy				
Machinery and transport equipment		4	4	184
Electronics/EDP	1	4	2	69
Electrical equipment		2		31
Light industry:				
- textiles and leather		4		476
- furniture				
- consumer goods		1		145
Other branches:				
- construction and building materials	1			
- hotels				
- tourism		1		
- banking and financial services		4		
- publishing				
- trading		4		80
- other services	5	21	1	315

(a) In number of companies operating in a given sector

Source: Ostwirtschaftsreport 11/10/1985, pp. 401-403; M. Malecki-Spőlki z udzialem kapitalu zagranicznego ... op.cit.

Table 6.9. Index of Comparative Advantages of Individual CMEA Countries (a) by branch (as of 1984)

Country	Food and agriculture 1	Chemicals 2	Metallurgy 3	Transport Equipment 4	Machine Tools 5	Mechanical Engineering 6	Electronics 7	Electrical equipment 8	Light Industry 9	Other (construction, services) 10
Bulgaria	25/9.1	0.0/9.1	0.9/4.5	0.0/6.8	0.0/4.5	25.0/25.0	0.0/11.4	0.0/6.8	50.0/20.4	0.0/2.3
Hungary	8.0/8.9	20.0/18.2	4.0/3.1	2.0/8.6	0.0/4.8	6.8/11.7	8.0/7.6	14.0/14.1	20.0/15.8	18.0/7.2
Poland	40.0/4.2	0.0/17.7	0.0/12.5	20.0/15.6	0.0/2.1	20.0/18.7	0.0/12.5	0.0/5.2	0.0/6.2	20.0/5.2
Rumania	4.5/2.6	40.9/34.2	4.5/10.5	26.4/34.2	0.0/0.0	0.0/0.0	4.5/5.3	0.0/2.6	0.0/5.3	9.1/5.3
Czecho-slovakia	0.0/10.1	20.0/8.7	0.0/0.0	0.0/11.6	40.0/16.0	0.0/21.7	0.0/2.9	20.0/14.5	20.0/11.6	0.0/2.9
USSR	12.5/4.9	18.7/35.0	12.5/9.9	0.0/7.8	12.5/5.3	25.0/13.6	0.0/5.8	0.0/3.3	6.2/3.3	12.5/8.6
Subtotal:	10.2/6.8	21.3/22.5	4.6/6.5	9.3/11.3	4.6/5.3	9.3/14.4	5.5/7.4	7.4/8.4	14.8/10.8	13.0/6.5
Yugoslavia	11.4/7.8	24.4/16.7	6.9/5.3	16.8/20.4	1.5/1.6	8.4/12.2	5.3/6.5	4.6/9.0	13.7/12.6	6.9/7.8
Total:	10.9/7.0	23.0/21.1	5.9/6.2	13.4/13.4	2.9/4.6	8.8/13.9	5.4/7.2	5.9/8.5	14.2/11.2	9.6/6.8

(a) calculated as $\dfrac{FDI_j^i}{FDI_j} : \dfrac{JCC_j^i}{JCC_j}$ where

FDI   denotes number of joint-ventures
JCC   denotes number of industrial cooperation contracts
j   country
i   branch

underlined figures - revealed comparative advantages

Source: Calculated from ECE, TRADE/R.487, October 19, 1984

in machine tools, mechanical engineering, electronics and electrical equipment. If we include Yugoslavia, our picture still does not change substantially.

This initially data, on the other hand, seem to confirm the validity of the hypothesis which states that one can expect less technological processes (ie. less food, chemicals and metallurgy) than technological products in first contracts, in comparison to other forms of technology transfer. However, if we include the data on Yugoslavia, this contention is no longer valid.

5. FDI and Technology Transfer to CMEA: Some Conclusions

As discussed earlier, FDI are not yet an important vehicle for foreign capital inflow to CMEA or for intensified technology inflow. This is quite contrary to what available theories seem to stipulate. How can we then explain this apparent paradox?

It appears, that at least three different, albeit not mutually exclusive, hypotheses can be formulated. Let us call the first one the "half-pregnant" hypothesis, the second one the "assymetrical cost-benefit configuration" hypothesis and the third, the "Wild West" hypothesis.

According to the "half-pregnant" hypothesis, a major cause for the apparent unattractiveness of investments in CMEA countries for Western investors is the limitation of foreign ownership to the minority share (49:51 principle). The limitation of property rights and the resulting limitation of controlling rights, makes the utilization of the FDI channel for Western companies as impossible as it is to be "half pregnant". It implies, therefore, that if the said limitations were aborted, one would witness a substantial increase in the inflow of capital and technology. Both empirical data and opinions expressed by a number of Western investors seem to contradict this hypothesis. As a matter of fact, from the six CMEA countries which permit Western investments locally only Rumania, the USSR and Czechoslovakia exclude Western majority participation. At the same time, it seems that Rumania was able to attract the most technology intensive investments (though their number is small). On the other hand, the possibility of Western investors holding the majority share in Hungary, Bulgaria and Poland has not materialized as predicted.

Also, the opinions expressed by Western investors on their hesitant attitude towards joint-ventures with CMEA countries, do not stress the importance of the property limitations, but rather the importance of the costs involved and the anticipated benefits. This brings us directly to the second hypothesis, the <u>asymetrical cost-benefit configuration</u>. Its basic contention is that currently prospective investors are faced with the high transaction costs of FDI undertakings, which are not compensated by the respective profits.

The high transaction costs are primarily the result of the overall systemic differences between the market and planned economies. Prospecitve investors usually find it very difficult to learn how to operate within planned economies, which are, by and large, an alien environment for most of them. It is not surprising, therefore, that the future investors are predominantly those Western companies which have acquired relevant knowledge about the planned economies in the course of earlier arm-length contracts with CMEA countries.

Other reasons for high transaction costs result from the adopted national legislation. This is frequently unclear, the relevant procedures over-bureaucratized, time consuming and the final decisions often unpredictable. In Hungary, for example, a good part of the joint-ventures required 5-6 years of preparatory work before they were put into operation. Similar experiences were recorded in Rumania as well as in Poland. The relevant regulations are also often viewed as very restrictive in the sense that they limit investors', freedom of action in day-to-day business, i.e. in setting up salaries, using company cars, earned foreign exchange and the like.

A third, but equally important element of transaction costs - also specifically related to CMEA - is connected with political and strategic limitations imposed both on capital and technology flows to CMEA by Western countries.

What is particularly important in this context is the fact that both system-conditioned and regionally-conditioned transaction costs (first and last categories) are extra costs for prospective Western investors, pertinent exclusiveley in the case of their capital involvement in CMEA and largely absent in the case of other investments. This means, that as a compensation particularly strong incentives must be offered, which would enable the attainment of a high rate of profit. This is, however, apparently not the case and hence there is a lack of motivations for substantial investments from the West. The aforementioned hypothesis also implies that the theoretical explanations of FDI and technology transfer concentrate too much of their attention on property rights and communication issues and apparently underestimate other relevant elements. Therefore, they cannot provide a proper explanation for East-West relations in this area.

Finally, it could also be claimed that the particular behavior of Western investors vis-à-vis CMEA countries is a result of the fact that they are currently only at the beginning of the relativ ly new development which - like the development of the Wild West - attracts only a certain type of investor, a sort of adventurer, hunting for a quick and speculative profit. Hence, the name, The Wild West hypothesis. This means that the bulk of investments is based on a catch and go strategy and thus they must be small and non-capital intensive. With the passing time and the "normalization" of this new phenomenon, the size and composition of Western FDI's will change accordingly and only then will the theories considered above suit the East-West reality.

Of course, the three explanations outlined above may be valid simultaneously or they may be, to some extent, mutually exclusive. There might be yet another explanation which we could not detect. To find it, however, would require much additional research.

References and Footnotes

1   K. Poznanski - The invironment for technological change in centrally planned economies, Dept. of Economics, Rensselear Polytechnic Institute, mimeo, October 1984

2   O. Williamson - The modern corporation: origins, evolution, attributes, in: Journal of Economic Literature, Dec. 19, 1981, pp. 1537-1568; P. Buckley, M. Casson - The future of the multinational enterprise, New York 1976; R. Findlay - Relative backwardness, direct foreign investment and the transfer of technology; a simple dynamic model, in: The Quarterly Journal of Economics, Vol. XCII, Feb. 1978, No. 1, pp. 11-16

3   O. Williamson, The modern corporation ..., op.cit., p. 1537

4   J. Harders, Arzneimittelforschung und Industrieorganisation - DDR und Ungarn im Vergleich, Ökonomische Studien, Band 37, Stuttgart/New York, 1985, pp. 40-50

5   S. Magee, Information and multinational corporation: an appropriability theory of direct foreign investment, in: J.N. Bhagwati (ed.), The new international economic order: The North/South debate, Cambridge, MIT Press 1977, pp. 317-340

6   R. Findlay, Relative backwardness ..., op.cit., pp. 4-5

7   R. Findlay, Relative bachwardness ..., op.cit., p. 5

8   J. Brada, Industry structure and East-West technology transfer: a case study of the pharmaceutical industry, in: ACES, No. 22, spring 1980, pp. 31-59; C.C. Coughlin, The relationship between foreign ownership and technology transfer, Journal of Comparative Economics, 7, 1983, pp. 400

9   R. Clapham, Zum Einfluß der Wirtschaftsordnung auf den internationalen Technologietransfer, ORDO, Band 25, 1974, p. 211

10 E. Mansfield, A. Romeo, W. Wagner, Foreign trade and US research and development, The Review of Economics and Statistics, Vol. 61, 1979, pp. 49-57

11 S. Globerman, Technological diffusion in the Canadian tools and die industry, The Review of Economics and Statistics, Vol. 57, 1975, pp. 428-434

12 E.K.Y. Chen, Multinational corporations and technology diffusion in Hong Kong manufacturing, Applied Economics, Vol. 15, 1983, pp. 309-321

13 C.H. McMillan - The foreign investment activity of the Comecon countries: actors and strategies, University of Reading, Discussion Papers in International Investments and Business Studies, No. 73, June 1983, p. 1; C.H. McMillan - Multinationals from the second world, Macmillan Press, London 1987, pp. 38-39

14 M. Lebkowski, J. Monkiewicz, Western direct investment in centrally planned economies, Journal of World Trade Law, Vol. 20, No. 6, November/December 1986, p. 627

15 M. Malecki, Spólki z udzialem kapitalu zagranicznego w europejskich krajach socjalistycznych, Warszawa 1987, mimeo, p. 27

16 T. Kaminski, Uwarunkowania kooperacji kapitalowo-przemyslowej z zagranica, PPIPH, Warszawa 1987, mimeo, p. 7

17 M. Malecki, Spólki z udzialem ... op.cit., p. 31

Part III
Dynamics of the CMEA Technological Position During the 1970s and 1980s

Chapter 7
Technology Gap

Conceptual and Statistical Problems

1. Introduction

Discussions of inter-country differences in the level of technological development have a long tradition. By and large, however, in the last 20 or so years they have been more inspired and attended by journalists and politicians than by researchers. As such, the issue has frequently been accompanied by intense emotions and has been highly politicized. Here, it is sufficient to mention the wave of publications, declarations and suggested political actions resulting from the publication of the famous book by the French journalist J.J. Servan-Schreiber, entitled "American challenge". The book presented a dramatic view of the declining West European technological position vis-à-vis the United States and demanded an overall mobilization to counteract American technological invasion (1). In response to the discussion, Amintore Fanfani, the Italian minister of foreign affairs at that time, put forward a proposal for the technological Marshall Plan and the British Prime Minister, Harold Wilson, demanded that a European technological community should be established. Thereafter, however, with the publication of the OECD research report "Gaps in technology", and some other more well founded statistical analyses which pointed out the considerable over-dramatization of the issue, the public interest ceased and the topic largely disappeared from the scene.

The topic arose again in the late 1970s with the advances in the area of micro-electronics and telecommunications and the fear of American-Japanese domination.

The question of the East-West technological gap has been given clearly much less attention. This is largely due to the belief that Eastern Europe does not pose a real technological threat to the West's domination in this area. In the majority of relevant literature the said question is discussed very briefly and with some astonishingly crude methods (2). Sound statistical analyses are very rare.

2. The Concept of the Technology Gap

Inter-country comparisons of technological levels are faced with two great obstacles: that of methodology and that of statistics. Both seem to be far from being resolved. Interestingly enough, apparently much more attention has been devoted in the past to statistical problems than to the conceptualization of the phenomenon itself. Only since recently, have both the OECD and the US Office for Technology Assessment initiated special research projects to clarify the concepts in question (3).

The concept of the technology gap was formulated for the first time in the neo-technology models of international trade back in the 1960s. In his famous paper on the relationship between international trade and technical change, M.V. Posner introduced the idea of the so-called imitation lag to identify the inter-country differences in technological performance (4). The imitation lag was measured by the time difference between the introduction of the new technology in the innovation country and the moment of its adoption in the imitating country.

Thus, the differences among the countries were to be measured on a time scale and the existing gaps expressed in time units. The implicit assumption of the suggested methodology was that technological change is predominantly of an imitative nature and that lags and leads in different areas of technology could somehow be summed up together. Crude as it is, this concept has been used fairly frequently, particularly in public discussions, due to its simplicity.

A great leap forwards in the methodology was subsequently offered by the much publicized OECD study on the technology gap between Europe and the US. In contrast to M.V. Posner, it viewed the technology gap as a multidimensional phenomenon which was reflected in:

"A. Differences in the development of national scientific and technological capabilities;

B. Differences in member countries' performance in technological innovation;

C. The economic effects of A and B, including the effects of international economic and technological exchange" (5).

The first element was supposed to represent the ability of different countries to create the scientific and technical preconditions for the subsequent technological change. It was measured in the said study with the help of analyses of the scientific and technical information system, scientific and technical manpower resources and the registered research and development effort.

The second element of the concept described the actual utilization of the available resources (materialization of technological change). It was further split into two components: performance in originating innovations (measured by the data on the location of some 140 significant innovations since 1945; receipts for patents and know-how and trade performance in R+D dependent products) and performance in diffusung innovations (level and rate of increase in the use of significant new products and production processes, rates of increase in total factor productivity).

The third and last component of the said concept was meant to measure the impact of technological change on both internal economic activity of the country as well as its external economic relations.

The OECD methodology was further refined by H. Majer, who suggested to measure inter-country differences in technology according to the dynamic perception of technological change, viewed as a sequence of research, development and implementation (6). This approach seems to offer the best possibilities for a policy-oriented analysis, as it can be detected in which stage of the overall process of the technological change eventual differences exist. It also enables a differentiation of possible policy measures according to the costs of their implementation and the necessary time frame.

3. Dimensions of National Technological Level

Taking the concept of H. Majer as a point of departure, it is suggested to analyze the national (regional) technological level as a system composed of three inter-related components: embryonic technology, potential technology and applied technology (technology in actual use).

In this concept embryonic technology represents the available resources for future technological development. It is principally connected with the scientific research process and is accordingly measured with the indicators characterizing this area of human activity. The current level of embryonic technology determines to a considerable extent the long-run technological position of a given country. The time lag incurred is principally dependent on the speed of building up the relevant scientific and research infrastructure which, as a rule, is fairly long. This is particularly true if more broad areas of technology are concerned. Improvements in the embryonic technology level thus require much time. On the other hand, they are relatively cheap in economic terms, in comparison to the other components of the system in question.

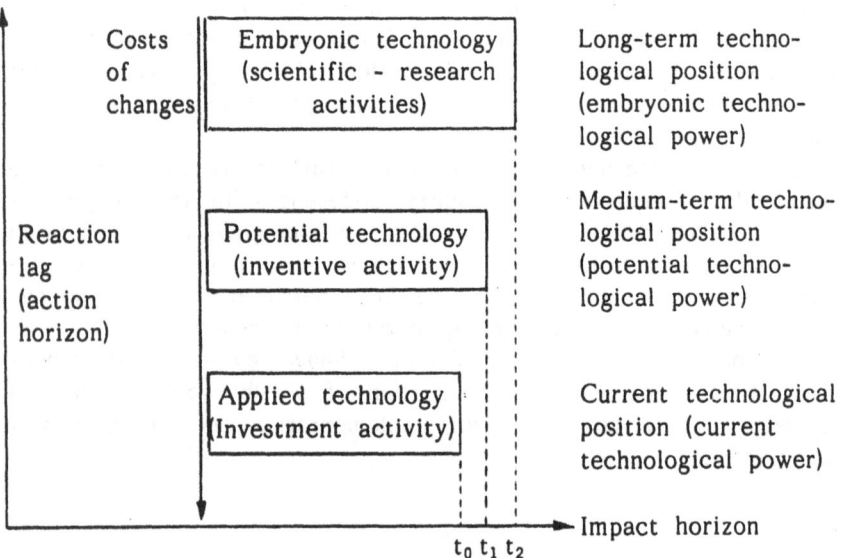

Fig. 7.1. Dimensions of National Technological Level/Technology Gap

Potential technology is defined as the already existing stock of available technologies, which changes over time and is primarily determined by the rate and size of the inventive activities. The level of potential technology determines the level of applied technology, on a medium-term basis.

Finally, applied technology is defined here as the technology in actual use, as reflected in a country's actual economic performance. Its changes over time are basically determined by the rate and size of the investment activity. The basic elements of the concept are presented in Fig. 7.1.

4. Statistical Indicators

The multi-dimensionality of the proposed approach necessitates the application of the variables pertinent to the identified process or technological development. Thus one has to apply three groups of indicators:

1. describing the level and structure of the scientific research activity;
2. describing the level and structure of the technical creativity;
3. describing the technological level and structure of the production activity.

The first group of indicators is meant to express the embryonic technological power. To analyze this in an international perspective, the following measures could be applied:

a) international quotation index
b) internationally renowned prizes and honors received
c) research manpower resources and their reproduction rate
d) qualitative analysis of the scientific level by experts.

In view of the lack of available data, however, some proxy measures, basically describing the national research efforts, are frequently used instead. These could include inter alia:

a) research and development outlays
b) sectoral and branch structure of R+D outlays
c) educational expenses
d) number of students enrolled, etc.

Thus, as one can see, they represent a type of input (as opposed to output) characteristics, and as such they neglect the efficiency aspect.

As far as the potential technological power is concerned, particular attention should be devoted to the patent statistics. These have been used fairly often in the past, both for the measurement of the technological level and for the detection of basic directions of technological change.

This method was somewhat abandoned in the 1970s, but it has recently gained more popularity and numerous attempts have been undertaken to make use of it for the analysis of the technological development of individual countries and sectors (7).

The use of patent statistics has several advantages. These include inter alia:

1. An easy access to statistical material which has already been collected in an organized manner for many years.
2. It is the only continous statistical information on the results of the national inventive activities. As estimated by the West German Institute for Economic Research (IFO) in Munich, around half of all inventions are normally patented (8).
3. The said statistics cover, in principle, the main inventive achievements of individual countries (which is a direct effect of the definition of a patent and rules for granting it).
4. It is expressed in physical units and classified according to the universally accepted international patent classification (IPC), which allows for easy cross-country comparisons.

At the same time, however, the use of patent statistics has some significant disadvantages, which principally incude:

1. The differentiated patent laws in the individual countries and, hence, the somewhat different criteria for granting patents (the differences, however, are generally not so large).
2. The economic function of the patent, which means that the patenting performance of the individual countries reflects the nature of their economic and particularly of the foreign trade policies and not only, or principally, their national inventive performance.
3. The purely quantitative character of the said statistics, with the effect that no information on the technological or economic importance of the patented inventions is available.

4. The differentiated significance placed on the patent protection for various classes of inventions (more important for product than for process innovations, in which case secrecy may be a preferred solution), which changes the actual structure of the inventive results.
5. The differentiated patenting propensity of individual industrial branches due mainly to the differing degree of imitation risk (danger of possible copying) and the different speed of technical change.

Hence, when using patent statistics for depicting the potential technological power of a given country, one should be aware of the inherent limitations and formulate the conclusions accordingly.

Patent statistics may characterize the overall state of the potential technological power (number of local patents registered), its changes (dynamics of a current patenting activity), international competitiveness (intensity of foreign patenting activity), as well as its internal structure, according to the main areas of technology. Also important is the fact that the said characteristics are given according to the lines of the types of technological activities and not according to the institutional structure of the industry (branches and sectors), which is the case with most of the other available indicators. This enables a type of a cross-sectional analysis, which is otherwise impossible. This last feature of the patent statistics also has its drawbacks. The international patent classification is based on completely different principles than industrial and trade classifications and, hence, straightforward comparisons between the three are nearly impossible. With some additional effort, however, this difficulty can be overcome.

Although significant, patent statistics are not the only means for measuring the potential technological power of a nation. Statistics on trade in technology can also be used. It should be noted, however, that the said statistics do not represent "pure" potential technology, as it partly also includes the technology actually in use. Therefore, it is a type of bridge linking the potential with the applied technologies.

The first direction of analysis which may be pursued in this regard is the utilization of date on the level and qualitative structure of technology exports (9). It is assumed that the technology export reflects a country's level of technological development both through its size and particularly through its internal structure. This reasoning is based on the observation that the technological development of any country goes through certain stages (see chapter 1).

For each stage of the technological development, specific forms of technology export could be assigned. Thus, the first stage could be charaterized by an increase in the export of simple industrial services, basically related to the manning, maintenance and repair of industrial plants and equipment. At the second stage, the export of design and consulting services emerges; at the third stage, construction and engineering services arise; at the fourth, the export of turn-key plants takes place and at the fifth, there is evidence of the export of licences. Cross-country comparisons of the registered structure of technological exports indicate the position a given country enjoys among other nations (see Fig. 7.2.).

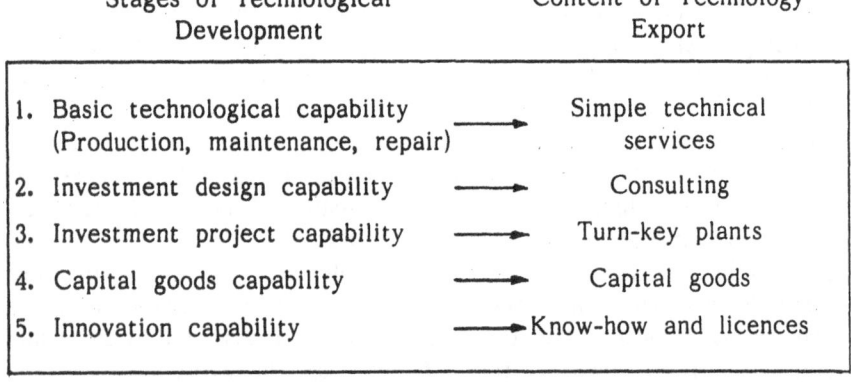

Stages of Technological Development	Content of Technology Export
1. Basic technological capability (Production, maintenance, repair)	Simple technical services
2. Investment design capability	Consulting
3. Investment project capability	Turn-key plants
4. Capital goods capability	Capital goods
5. Innovation capability	Know-how and licences

Fig. 7.2. Stages of Technological Development and Contents of Technology Export

Additional information which could be used here refers to the registered sources of comparative advantages (prices, delivery time, level of technology, etc.), the industrial structure of the exports (leading, secondary or sunset industries), the geographical (demanding or undemanding markets) destination, as well as the institutional destination (leading actors or market outsiders).

Apart from the utilization of statistics on technology exports, it may be also advisable to make use of the data on technology imports. A theoretical base for such a proposal results from the assumption that the intensity of the technology import and its industrial structure reflect the size and structure of the technological demand of the national economy which cannot be satisfied with local capabilities.

As far as the level of applied technology (technology in actual use) is concerned, it can be characterized mainly with indicators coming from the area of production and foreign trade. Indicators originating from production are apparently now in more wide use. They include such synthetic measures as productivity, resource intensity, energy intensity as well as a spectrum of natural indices such as the number of computers installed, the structure of manufacturing technologies used, etc.

As far as indicators related to foreign trade are concerned, two seem to merit our attention. The first is related to the analysis of the technological structure of foreign trade. The underlying hypothesis assumes that the higher the technological level of a country and the higher its share of technology intensive products (TIP) in its exports, the better its trade balance in this product group. Such analyses have been fairly frequently performed in recent times and, hence, there is no need to dwell more on this here (10).

The second proposition is based on the use of unit price or kilogram price indices, i.e. prices received (paid) in relation to the weight exported (imported). It is assumed that the higher this ratio is in comparison to other countries, then the higher the level of applied technology of the country in question. The principal methodological problem which emerges here is the question what other factors, apart from the technological level of the products, influence the said variable. Generally, the following sources of possible disturbances exist (11):

1. High kilogram prices could be a reflection of the particularly expensive inputs used (e.g. platinum). This reservation may be considered as important only at a very limited level of aggregation (individual products). It is basically unimportant when broad product groups are analyzed, and when changes in the level of unit prices are in the forefront instead of their actual level.

2. High unit prices may be a reflection of the individual nature of the product (e.g. art products). Again, this reservation seems to be unimportant in the case of larger aggregates analyzed.

3. Low unit prices not necessarily reflect the low level of technology embodied, but the fact that the products in question are simply physically overdimensioned. Here one could argue, however, that this is an indication of the overall technological backwardness of the country either in design or in material manufacturing.
4. The level of unit prices can also be influenced by some peculiarities in export policies (e.g. export promotion via low prices) or in import, by difficult access to some products (e.g. embargo).
5. Finally, differences in unit prices may reflect differences in the scale of production (12).

All of these reservations appear to relate to some secondary effects and thus do not undermine the usefulness of the aforementioned measure for our purposes.

Summarizing the above, one can argue that the proper assessment of a country's technological position in an international perspective should be based on a multidimensional and three-step approach (Fig. 7.3.).

Levels of Measurement	Indicators
Embryonic technology	Input indicators: R+D outlays, student enrollment, R+D manpower Output indicators: quotation index, prices, honors
Potential technology	Patent statistics Trade in technology statistics
Applied technology	Production related indicators Foreign trade related indicators

Fig. 7.3. Measurement of the Technological Level of a Country (Region)

The concept outlined does not allow the position of a given country to be defined with a high degree of accuracy, however, due to its multidimensionality, it registers the phenomenon in the most objective way possible.

References and Footnotes

1   J.J. Servan-Schreiber - Die amerikanische Herausforderung, Hamburg 1968
2   In a recent voluminous OECD study, "East-West technology transfer. The trade and economic aspects", Paris 1986, the technology gap is discussed on two pages.
3   See: Science indicators 1982. An analysis of the state of US science, engineering and technology, Report of the National Science Board, New York 1983
4   M.V. Posner - International trade and technical change, Oxford Economic Papers, Vol. 13, 1961, pp. 323-341
5   Gaps in technology. General report, OECD, Paris 1968
6   H. Majer - Die technologische Lücke zwischen der Bundesrepublik Deutschland und den Vereinigten Staaten von Amerika, J.C.B. Mohr, Tübingen, 1973
7   See: D. Schiftel, C. Kitti - Rates of invention: international patent comparisons, Research Policy, no. 7/1978, pp. 323-340; M. Macioti - The power and the glory: a note on patents and scientific authors, Research Policy, no. 9/1980, pp. 104-114; K. Pavitt - Patent statistics as indicators of innovation activities: possibilities and problems, SPRU, University of Sussex, December 18, 1983, mimeo, pp. 42
8   U. Chr. Täger - Untersuchung der Aussagefähigkeit von Patentstatistiken hinsichtlich technologischer Entwicklungen, IFO, Studien zur Industriewirtschaft, Nr. 17, München 1979, p. 49
9   This line of analysis has been developed in particular by the World Bank in its studies on the technological build-up of developing countries. See, for example, L.E. Westphal, et al. - Exports of capital goods and related services from the Republic of Korea, The World Bank, October 1983, mimeo
10 See for example: O. Börnsen, et al. - Der Technologietransfer zwischen den USA und der Bundesrepublik, Kieler Studien, Nr. 192, Mohr, Tübingen 1985

11 R. Amman, J. Slama - The organic chemical industry of the USSR: a case study in the measurement of comparative technological sophistication by means of kilogram-prices, Research Policy, no. 5/1976, pp. 302-326
12 Measuring the changing technological level of international trade flows by means of unit values: some empirical findings, ECE, UN, Trade IR.463, August 12, 1983, p. 2

Chapter 8
Evolution of the CMEA Technological Position During the 1970s and the 1980s

1. Introduction

It was argued in the preceding chapter that the technological level of a country, or a country grouping, should be analyzed from three differnt, though inter-related dimensions: embryonic technology, potential technology and applied technology. The current international competitiveness of a country is determined by the state of art of its applied technology and changes occuring here affect the international competitiveness. The state of art of applied technology is, on the other hand, - especially in autarchic or semi-autarchic economic system - basically a result of the state of art of the country's potential technology, which, in turn, is dependent on the level and dynamics of the embryonic technology.

The said relationship is not mechanistic but definitely dialectic. This means that there is specific feedback among the elements mentioned. Thus, for example, the level of applied technology is, on the one hand, determined by the state of potential technology and, on the other, it is one of the forces shaping the level of applied technology. It also means that there is no rule of strict determinism among the elements in question.

In the following, we shall try to implement the said methodology in assessing the CMEA technological position during the last two decades. We shall concentrate our attention on potential and applied technologies, as precisely here the statistical base is sufficient enough to allow for a sound analysis.

2. CMEA Embryonic Technology in the 1970s and 1980s

It is generally acknowledged by experts on CMEA countries that as far as embryonic technology is concerned, the countries in question meet basically Western standards or even exceed them. This conviction is based foremostly on expert assessments as no attempt at a sound statistical analysis has been made so far. The principal reason is the apparent lack of comparable statistical data. This finally results in the fact that the existing analytical approaches make use predominantly of the relevant figures on research and development activities.

Table 8.1. Expenditures on R+D as Percentage of GNP or GDP

	1967	1975	1982
CMEA			
1. Bulgaria	1.4	2.2	2.8
2. Czechoslovakia	3.6[a]	3.9	4.0
3. GDR	.	3.0	4.4[a]
4. Hungary	2.3	3.4	2.6
5. Poland	1.8	-	2.2
6. USSR	3.7	4.8	4.7
OECD			
1. Austria	0.6[c]	0.9	1.2
2. Belgium	0.7	1.3	1.4[b]
3. France	2.2	1.8	2.0
4. GFR	1.9	2.2	2.5
5. Italy	0.7	0.9	1.0
6. USA	2.7	2.4	2.5

a - 1981,   b - 1979,   c - 1970

Source: Compiled from ECE Sc.TECH./R.175/Add. 3, August 22, 1985, pp. 2-3

According to the said figures, CMEA countries represent over 1/4 of the world's total R+D expenditures and their share during the 1970s was fairly constant. Compared with the OECD countries, however, they spend much more on R+D in terms of their national income (see Table 8.1.). This is especially true with respect to the Soviet Union, the GDR and Czechoslovakia.

The position of the CMEA countries is even more pronounced when the figures on R+D employment are taken into account. Soviet scientific manpower alone is estimated at 1.4 million persons (0.8 million in 1970), which is about twice as much as the relevant figure for the United States (1). According to some estimates, in 1980 CMEA countries employed over 60% of the world's total R+D personnel (2).

There are no indications of any dramatic shifts in the said area during the 1970s and, hence, we can assume that the CMEA's strong position remained unchanged. Available data indicate that basically the internal ranking of individual CMEA countries has also remained unaltered. The Soviet Union, with its vast research resources and large projects, is clearly the principal actor in the game, followed by the German Democratic Republic, Czechoslovakia and Poland (see Table 8.2.).

Since the beginning of the 1980s, Poland has lost much of its previous standing due to the dramatic structural crisis and, thus, decreased both its governmental and company outlays for innovation and investment activities. Recently, it has started to slowly regain its former position.

3. CMEA Potential Technology in the 1970s. Trends and Patterns

As argued before, principal information on the state of the potential technology can be derived from patent statistics. It can be used both to characterize the internal and external patenting power of a country. The internal patenting power illustrates the size and dynamics of national inventive activities and, thus, it is limited to a purely national dimension of the said phenomenon. The external patenting power, on the other hand, illustrates the position of a country within international patent circles. Hence, it characterizes a foreign dimension of the phenomenon.

The simplest indicator of the internal patenting power is the registered stock of national patents in force. It is a cumulative figure, comprising the results of national inventive activity of the last 7-10 years (this is a standard time period of patent holding). Some may be in actual use and some may still be waiting for the eventual application.

The problem with the said indicator is that the necessary figures are largely unavailable, as only a few contries supply relevant data. From the fragmentary data accessible, however, it seems that CMEA countries enjoy a relatively good standing in the international community in this respect. This is further confirmed by the figures on patents granted to nationals in individual years which measures the overall dynamics of inventive performance (see Table 8.3.).

As indicated by the data provided in Table 8.3., CMEA countries demonstrated substantial internal patenting power throughout the 1970s and their share in patents granted worldwide climbed from ca. 30% of the world total in the early 1970s, to nearly 50% of the world total at the beginning of the 1980s, to decrease thereafter to 40% in 1984. In 1984, four of the CMEA countries: the USSR, Czechoslovakia, the GDR and Poland, were among the 10 leading countries in the world, with regard to the number of patents granted to nationals. A particular rapid increase in the internal patenting power in this period was registered in the Soviet Union, whereas the remaining 6 CMEA countries showed differentiated performance during this time. Polish patenting performance clearly declined, that of Rumania and Hungary was more or less constant, and the GDR's increased. Industrialized western countries, on the other hand, experienced a substantial weakening of their patenting performance in the 1970s, both in relative as well as absolute terms (with a few exceptions), but improved their performance thereafter. Assuming similar technical value of the patents granted in individual countries, as well as a similar propensity to patent, this could indicate a future shift in the potential technological power among the countries concerned and, consequently, the possibility of a future shift in applied technology (given the same efficiency in the innovation processes).

The relatively good standing of the CMEA countries with respect to internal patenting power is also confirmed by data on patent filing per capita. Again, four of the CMEA countries: the USSAR, the GDR, Czechoslovakia and Bulgaria, are among the ten best perfomers in the world, together with Japan, the FRG, Switzerland, Sweden, Australia and New Zealand (see Table 8.4.).

Table 8.2. Indicators of Science and Technology Potentials of the CMEA Countries

Country	Share of R+D in National Income (%)			Overall Employment in S and T (000)			Scientists and Engineers engaged in R+D (000)			No. of Independent R+D Establishments		
	1970	1975	1980	1970	1975	1980	1970	1975	1980	1970	1975	1980
Bulgaria	2.1	2.3	2.6	46.7	60.5	62.4	12.7	18.4	22.6	144	317	276
Czechoslovakia	3.4	3.8	3.8	165.0	176.0	193.7	37.9	52.6	69.8	145	131	132
GDR	3.9	4.4	4.5	143.0	1603	186.0	30.0	50.0	70.0	1800[a]	1380	–
Hungary	2.8	3.5	3.5	70.0	81.3	85.0	23.7	34.8	38.5	131	128	126
Poland	2.0	1.8	2.0	256.7	288.0	357.4	39.1	60.5	134.5	245	410	201[b]
Rumania	0.9	1.3	1.4	58.0	84.3	110.0	14.0	–	–	158	143[c]	–
USSR	4.0	4.7	4.7	3238.0	3792.0	4350.0	927.7	1223.4	1373.3	2458.0	3531[c]	–

a) including design offices
b) as of 1983
c) as of 1974

Source: S.I. Stiepanianko - SEV: Miezhdunarodnoje sotrudnichestvo v oblasti nauki i tiekhniki, Moskva 1985, p. 63

Table 8.3. Internal Patenting Power of the CMEA Region (patents granted to nationals at home)
(a)

Country	1970 Number	1975 Number	1975 % of World Total	1978 Number	1978 % of World Total	1981 Number	1981 % of World Total	1982 Number	1984 Number	1984 % of World Total
1. Bulgaria	534	752	0.4	1235	0.6	1417	0.6	1424	1372	0.6
2. Czechoslovakia	3224	4453	2.3	5261	2.7	5447	2.2	6113	6266	2.9
3. GDR	5308	3663	1.9	3305	1.7	5713	2.3	4125	9538	4.4
4. Hungary	481	634	0.3	602	0.3	764	0.3	760	1450	0.7
5. Poland	2180	6659	3.5	4656	2.4	4693	1.9	3607	3532	1.6
6. Rumania	1440	1001	0.5	1743	0.9	1228	0.5	1581	1660	0.8
7. USSR	43803	39740	20.7	51839	26.2	96537	39.6	89304	62744	29.0
1-7 total	43803	56902	29.6	71735	36.3	115799	47.5	106914	86562	40.0
8. Australia	·	925	0.5	701	0.4	505	0.2	·	726	0.3
9. Austria	1198	1178	0.6	1180	0.6	1120	0.5	1158	1187	0.5
10. Belgium	1316	1034	0.5	975	0.5	788	0.3	896	834	0.4
11. Canada	1526	1280	0.7	1404	0.7	1369	0.6	1386	1427	0.7
12. France	8439	4962	2.6	8083	4.1	6855	2.8	7764	7651	3.5
14. Japan	21403	36992	19.3	37648	19.0	42080	17.3	40380	51690	23.9
15. Netherlands	284	419	0.2	432	0.2	395	0.2	618	757	0.4
16. U.K.	10343	9120	4.8	8464	4.3	6076	2.5	4686	4442	2.1
17. USA	47075	46603	24.3	40797	20.7	39225	16.1	33896	38364	17.7
8-17 total	·	111600	58.1	111447	56.3	104952	43.1	·	118480	54.8
World total	·	191962	100.0	197856	100.0	243554	100.0	·	216391	100.0

(a) including author's certificates

Source: Compiled from Industrial Property Statistics 1975-1984, WIPO

Table 8.4. CMEA Patenting Activity in an International Perspective

Country	Population (million)	National patent and utility models filling per 10,000 inhabitants	Rank
1. USSR	270.0	6.1	4
2. USA	229.3	2.7	14
3. Brazil	127.7	0.3	29
4. Japan	117.0	30.5	1
5. Mexico	71.3	0.1	30
6. FRG	61.7	8.9	2
7. Italy	57.4	1.1	25
8. U.K.	56.1	3.5	11
9. France	54.2	2.0	18
10. South Korea	41.1	2.2	16
11. Spain	37.9	2.0	17
12. Poland	36.3	2.4	15
13. Canada	24.4	0.7	28
14. Rumania	22.6	1.1	26
15. GDR	16.7	4.0	8
16. Czechoslovakia	15.4	4.9	6
17. Australia	15.0	4.4	7
18. Netherlands	14.3	1.323	
19. Hungary	10.7	1.5	22
20. Belgium	9.9	0.9	27
21. Greece	9.8	1.3	24
22. Bulgaria	8.9	3.7	9
23. Sweden	8.3	5.0	5
24. Austria	7.6	3.1	12
25. Switzerland	6.3	6.4	3
26. Denmark	5.1	1.9	19
27. Finland	4.8	2.8	13
28. Israel	4.1	1.6	21
29. Norway	4.1	1.8	20
30. New Zealand	3.1	3.7	10

Source: G. Kaspar - Die Bedeutung gewerblicher Schutzrechte für ein Stahlindustrieunternehmen im Hinblick auf seine Wirtschaftsbeziehungen mit RGW-Staaten, 2. Sonderseminar Technologietransfer und gewerblicher Rechtschutz im Ost-West-Handel, 10.-12. November 1986, Linz, p. 18

Table 8.5. Foreign Patenting Balance of CMEA and selected OECD Countries

Country	1970		1975		1980		1981		1982		1984	
	1	2	1	2	1	2	1	2	1	2	1	2
CMEA (x)												
Bulgaria	98	385	168	675	197	527	159	383	·	295	223	353
Czecho-slovakia	1512	1276	1039	2447	505	1853	456	1560	·	1636	537	1235
GDR	1844	2421	2430	2999	653	1371	817	1734	·	179	2488	2006
Poland	481	441	382	3356	407	1962	336	1409	·	856	195	653
Rumania	207	99	150	593	92	814	60	1276	·	920	32	925
Hungary	752	777	1260	1204	1351	1018	1335	840	·	1039	1802	1416
USSR	1725	1830	3011	2295	2725	1685	2343	1942	·	1792	2287	1114
Total CMEA	6619	7229	8440	13569	5774	9230	5506	9144	·	7717	7564	7702
Other Countries												
Austria	1859	7563	2000	1830	4745	1600	1600	4360	·	2903	1911	7378
Belgium	2260	15782	2113	12110	1752	5081	1417	3902	·	2930	1805	1930
France	15565	177758	15272	9358	13159	19622	11489	14622	·	16180	15131	16015
Netherlands	7717	2224	5880	3416	4016	2907	5421	2661	·	6035	6280	9500
Japan	10767	9415	18489	9736	20949	8974	19649	8824	·	·	29328	10110
FRG	32366	6501	39388	9213	33953	10362	30597	6892	·	8027	35050	10356
Switzerland	12691	13123	12225	9906	10620	4486	8378	6381	·	7712	9222	11626
USA	78286	17354	60926	25391	56240	24675	52572	26545	·	23993	56182	28837
U.K.	17179	30652	14489	31569	11937	18676	9974	16848	·	24904	11868	14425

(x) including author's certificates

Source: Industrial Property Statistics WIPO, Geneva, various years

The picture changes dramatically when foreign patenting performance of the CMEA countries is taken into account. The external patenting power of the said countries is extremely low both in terms of national patents filed abroad and of patents granted at home to foreigners (see Table 8.5.).

As indicated by the data in Table 8.5., the CMEA countries were granted abroad 5.5 to 8.5 thousand patents annually, with the maximum level reached in the mid 1970s. Thus, the foreign patenting performance of the said countries was, as a whole, around the level registered by the Netherlands alone and was considerably lower than that of larger OECD countries. If we look at the performance of the individual CMEA countries, we can see that in the first half of the 1970s the best results in this respect were registered by the GDR and the Soviet Union, and in the second half of the 1970s, by the Soviet Union and Hungary.

Foreign patenting activity may be characterized not only by the absolute number of patents filed (granted) abroad, but also by the ratio of patents granted abroad to patents granted to nationals at home. Let us call this ratio foreign patenting aggressiveness. It demonstrates the degree to which local inventions are given protection abroad, for whatever reason. The higher the said ratio, the higher the foreign patenting aggressiveness of a country. One should note, perhaps, that the patenting aggressiveness depends on the number of inventions scheduled for patenting abroad and on the density of protection sought (number of countries in which protection is sought).

The foreign patenting aggressiveness of CMEA countries was very low throughout the 1970s, in comparison to the level registered in OECD countries. The first were granted abroad much fewer patents than they registered at home (6 to 20 times less), whereas in case of OECD countries the proportions were quite different. They were granted abroad more patents than they actually registered at home. This is best demonstrated by the figures on the Swiss and West German performance. The said countries patented abroad 3 to 7 times more than at home. In general, the foreign patenting aggressiveness of the CMEA countries represented ca. 14% of the level registered by OECD member states in 1970, falling to slightly over 12.0% in 1984 (see Table 8.6.). A clear exception among the CMEA countries is Hungary, which seems to meet international standards in this respect.

Among the OECD countries, on the other hand, a peculiar situation exists in Japan, where foreign patenting aggressiveness is well below the level registered by other, both smaller and larger, OECD countries.

Table 8.6. Foreign Patenting Aggressiveness of CMEA and Selected OECD Countries

Country	1970	1980	1981	1984
CMEA	0.15	0.05	0.048	0.09
1. Bulgaria	0.18	0.15	0.11	0.16
2. Czechoslovakia	0.47	0.07	0.08	0.09
3. GDR	0.35	0.15	0.14	0.26
4. Poland	0.22	0.07	0.07	0.06
5. Rumania	0.14	0.08	0.05	0.02
6. Hungary	1.56	1.78	1.75	1.24
7. USSR	0.06	0.03	0.02	0.04
OECD				
8. Austria	1.56	1.49	1.43	1.61
9. Belgium	1.72	2.09	1.80	2.16
10. France	1.84	1.56	1.68	1.98
11. FRG	5.07	3.46	4.68	3.07
12. Japan	0.50	0.55	0.47	0.57
13. U.K.	1.66	2.31	1.64	2.67
14. USA	1.66	1.51	1.34	1.46
15. Switzerland	2.85	7.20	.	3.92

Source: calculated from WIPO, Industrial Property Statistics, 1970-1984

A pertinent question is how one can interpret this phenomenon. Why is the patenting aggressiveness of the CMEA countries so low? Why is Hungaria's position so high? What is wrong with Japan?

To answer these questions, one should note first of all that foreign patenting aggressiveness is apparently determined by two principal elements: the nature of the national inventions patented at home, and the nature of the foreign patenting policy of a country.

As far as the first element is concerned, one can assume that the higher the level of technical inventions patented at home and the more they relate to the strategically important areas of national production for export markets, the higher, ceteris paribus, the foreign patenting performance of a country.

As for the second element, the nature of the foreign patenting policy, it seems that it primarily reflects the type of foreign trade strategy pursued by a given country. An import substitution strategy results in a low patenting level abroad, whereas an export promotion policy ceteris paribus results in a relatively high level of foreign patenting. Obviously enough, there might be some other factors in operation, such as the commodity structure of the export (the more products requiring patent protection, the higher the patenting abroad), the destination of export flows (the higher the share of countries in which patent protection is irrelevant or unnecessary, the lower the level of patenting abroad), the availability of the financial means (especially hard currency) for the patenting activities abroad, etc.

It seems that in the case of CMEA countries, the main reasons for low foreign patenting aggressiveness are associated with the nature of their foreign patent policies, reflecting their import substitution foreign trade strategy. The second important reason is the fact that the overwhelming part of CMEA industrial export is destined to member countries, which have, by and large, substituted mutual competition (and thus undermined the economic functions of the patents) by administratively negotiated agreements on specialization and cooperation. In effect, intra-regional patenting occurs very seldom and is less important than extra-regional patenting. According to this hypothesis, the high level of the Hungarian foreign patenting aggressiveness is a result of its relatively strong export promotion, deviating from the import-substitution strategy of the other CMEA countries, and the relatively high share of industrial products in its export continuum to Western countries.

It does not appear that the low level of foreign patenting aggressiveness of the CMEA countries was based primarily on the low level of national inventions protected by national patents. Two arguments could be used to support this conviction. Firstly, in all of these countries patents (author's certificates) are granted only to inventions which represent an absolute world novelty and which must be confirmed during the patenting process. Second, the so-called "patent success coefficient" (measured as a ratio of patent applications filed abroad to patents granted) varies within international norms, around 0.6. Thus, the quality of CMEA applications seems to conform with that observed elsewhere.

Let us now consider another dimension of external patenting power, namely the degree to which a country attracts foreign patents, i.e. foreign patenting attractiveness.

Foreign patenting attractiveness may be measured both by the number of patents granted to foreigners (non-residents) and by the ratio of patents granted to non-residents to those granted to nationals (coefficient of foreign patenting penetration). In both cases, the position occupied by the CMEA countries is extremely poor (see Table 8.5. and Table 8.7.).

In 1982, for example, together they granted the same number of patents to non-residents as the Federal Republic of Germany alone.

Among the CMEA countries, the largest number of patents granted to foreigners was registered in the Soviet Union. All in all, however, it is significantly below the level registered in Austria or Belgium. The picture does not change significantly if the coefficient of foreign patenting penetration is calculated. Still, CMEA countries lag far behind the level observed in most of the OECD nations, with the exception of Japan and the US, in which case low values of the said coefficient were also detected. Among the CMEA member states, Hungary is clearly the leader, followed by Rumania, Bulgaria and the GDR.

The pertinent question is whether the low patenting attractiveness of the CMEA region reflects its poor technological endowment or whether its roots are located elsewhere.

The answer, however, is not as simple as one might think. Let us note, first of all, that foreign patenting attractiveness, measured by the coefficient of the foreign patenting penetration of individual OECD countries, is highly differentiated, i.e. the relevant coefficient in Japan is ca. 0.2; US - 0.7 and UK - 5.3. Let us also point out that in developing countries, and thus relatively

Table 8.7. Foreign Patenting Penetration of CMEA and other
Selected Countries

Country	1970	1975	1980	1982	1984
CMEA Total	0.17	0.23	0.08	0.07	0.09
1. Bulgaria	0.72	0.89	0.41	0.21	0.26
2. Czechoslovakia	0.40	0.55	0.27	0.27	0.20
3. GDR	0.46	0.82	0.31	0.29	0.21
4. Poland	0.20	0.50	0.34	0.24	0.18
5. Rumania	0.22	0.59	0.68	0.58	0.56
6. Hungary	1.61	1.90	1.34	0.95	0.98
7. USSR	0.06	0.05	0.02	0.02	0.02
OECD					
8. Austria	6.31	4.95	3.87	2.51	6.22
9. Belgium	7.83	11.71	6.07	3.27	2.31
10. France	2.12	1.89	2.33	2.08	2.09
11. FRG	1.02	1.01	1.05	0.97	0.91
12. Japan	0.44	0.26	0.21	.	0.20
13. U.K.	2.96	3.46	3.61	5.31	3.24
14. USA	0.37	0.54	0.66	0.71	0.75
15. Switzerland	2.95	6.72	3.04	4.03	4.95

Source: computed from WIPO, Industrial Property Statistics, 1970–
1984

poorly endowed in technology, the coefficient of foreign patenting
penetration is extremely high: 10 or more. This results, however,
not so much from the high intensity of foreign patent penetration,
but more from the low intensity of local patenting. The low values registered in the US and Japan seem to result from the ex-
act opposite situation: local patenting activity is so high that it

offsets the effects of the relatively high number of foreign patenting filings.

The high number of foreign filings may indicate both the technological attractiveness of a country (high level of locally developed technologies) and/or its market attractiveness (high expectations with respect to export possibilities). In the first instance, foreign patents are meant to play a defensive role; their function is to stop the technologically-based export offensive of a given country (region). In the second case, their role is clearly an offensive one; they are supposed to create a legal base for subsequent technology transfer operations or technology-based export to the markets in question. This occurs particularly frequently when a country undertakes vast development projects, i.e. when the demand for new technology is particularly high.

As far as the CMEA countries are concerned, defensive foreign filings, i.e. undertaken due to possible technological competition from their side, apparently play a negligible role. This is strongly justified especially by their import substitution strategy of foreign trade. Hence, changes in the coefficient of foreign patenting penetration are to be viewed as a result of a changing perception of their market attractiveness. As is well known, in all CMEA countries in the 1970s, there was a visible investment acceleration and parallel import acceleration, especially that of Western origin. Under these circumstances, one could logically expect an increase in their patent attractiveness during the said period. As the figures in Table 8.7. indicate, this was actually the case. In all CMEA countries, the coefficient of foreign patenting attractiveness rose from 1970-1975, falling subsequently with the deceleration of the investment programs of the said countries. The increase in the coefficient was certainly not a dramatic one, which could mean that the former level was basically exhausting the areas of technological advantage enjoyed by OECD countries. If this is true, then the registered data give a rather positive assessment of the potential technological power of the CMEA countries.

4. Technology-wise Distribution of the Patenting Power

Apart from measuring the overall potential technological power of a given nation (or region), patent statistics may be an interesting source of information on the directions of technological change, or more precisely, on the inter-country differences in the directions of technological change. The principal theoretical hypothesis underlying such reasoning is an assumption that the higher the patenting intensity of a country in certain areas of technology as compared to other countries, the larger the stock of available technologies and the stronger the technological position in these areas.

From a purely technical point of view, such comparisons are easy to carry out due to the existence of the universally accepted International Patent Classification. The only problem with classification is that it is based on completely different principles than the available industrial (ISIC) and trade classifications (SITC), which, in effect, makes it extremely difficult to draw conclusions with respect to sectors or branches of economic activity. With some risk of oversimplification, however, these difficulties can be overcome. Relevant figures on technology-wise distribution of CMEA patenting power are contained in Table 8.8.

From the eight main sections IPC is composed of, section A was dropped from the analysis as it contains primarily consumption-related inventions (human necessities) and thus, has no direct bearing on the technological level of a country. Some other sub-sections, due either to a lack of relevant figures or to their military orientation (like weapons and blasting), were also omitted. Therefore, the picture is far from being complete, though it still covers around 60% of all inventions patented worldwide.

An analysis of data contained in Table 8.8. reveals certain interesting patterns in the sectoral-wise patenting power of the CMEA countries.

First, it indicates a high degree of vulnerability in this power during the short time intervals analyzed. This might be interpreted as the lack of any clear cut technological trends or, more precisely, in the directions of foreign patenting of the said countries.

Second, on an average, the share of foreign patents granted to CMEA countries represents in all reported cases only a marginal fraction of worldwide patenting, normally oscillating around 3-4% of the world total. Thus, it means that on the whole the said countries can neither influence the overall pattern of worldwide patenting nor exert any meaningful impact on individual areas of technology.

Third, there are only a few obvious areas where CMEA countries have higher than average involvement in foreign patenting, the rest being spread out astonishingly even among the class in

Table 8.8. CMEA Technology-wise Distribution of Patenting Power 1975-1981 (share in patents granted abroad broken down by IPC classes, in % of total)

IPC Class Code	IPC Classes	Year	CMEA Total	CSSR	GDR	Hungary	USSR	Poland
1. B01-B09	Separating and mixing (physical or chemical processes or apparatus in general, crushing, pulverising, etc., separation, centrifugal apparatus or machines, spraying, etc.)	1975	3.3	0.5	0.7	0.4	1.7	·
		1978	8.1	0.3	3.2	1.5	2.1	1.0
		1981	3.4	0.2	0.3	1.6	1.6	0.7
2. B21-B23 and B24-B32	Shaping (mechanical metal working, casting, powder metallurgy, machine tools, grinding, polishing, hand tools, working or preserving wood, working cement, clay, stone, working of plastics, presses)	1975	5.1	0.4	0.9	0.1	3.7	·
		1978	6.2	0.2	2.2	1.6	1.7	0.5
		1981	2.2	0.1	0.4	0.1	1.2	0.5
3. B41-B44	Printing (printing, typewriters, bookbinding, bureau accessories, decorative arts)	1975	3.2	1.6	1.1	0.0	0.5	·
		1978	3.2	0.2	2.0	0.4	0.6	0.0
		1981	0.9	0.2	0.5	0.1	0.1	0.0
4. B60-B68	Transporting (vehicles in general, railways, ships, aircraft, aviation, conveying, packing, storing, lifting, liquid handling, etc.)	1975	2.7	0.3	1.3	0.1	1.0	·
		1978	5.0	0.3	2.3	1.5	0.8	0.0
		1981	1.3	0.1	0.2	0.2	0.4	0.4
5. C01-C14	Chemistry (inorganic chemistry, organic chemistry, organic macro-molecular compounds, dyes, paints, adhesives, biochemistry, skins, hides, etc.)	1975	2.7	0.3	0.7	0.9	0.8	·
		1978	6.0	0.3	2.6	1.9	1.0	0.2
		1981	2.6	0.1	2.0	1.3	0.7	0.2
6. C21-C30	Metallurgy (iron, alloys, working or treatment of metals, electolytic processes, crystal growth)	1975	3.6	0.1	0.4	0.4	2.7	·
		1978	7.4	0.3	3.2	1.7	1.7	0.5
		1981	2.6	0.2	0.1	0.5	1.6	0.3

No.	Code	Description	Year						
7.	D01-D07	Textiles and flexible materials (natural and artificial threads or fibers, yarns, weaving, braiding, sewing, treatment of textiles, ropes)	1975	4.8	1.8	1.8	0.1	1.1	·
			1978	5.9	0.8	2.4	1.0	1.4	0.3
			1981	2.5	0.5	1.0	0.3	0.7	0.0
8.	E01-E06	Building (construction of roads, railways and bridges, hydraulic engineering, water supply, building, locks, keys, doors, windows)	1975	2.5	0.1	0.9	0.3	1.2	·
			1978	3.3	0.2	0.9	1.6	0.5	0.1
			1981	0.9	0.0	0.2	0.3	0.2	2.0
9.	E21	Mining	1975	5.4	0.0	0.0	1.2	4.2	·
			1978	8.9	0.0	1.6	2.5	2.8	2.0
			1981	3.6	0.0	0.0	1.4	1.0	1.2
10.	F01-F17	Engines, pumps and engineering in general (machines and engines in general, combustion engines, pumps, fluid pressure actuators, engineering elements or units, etc.)	1975	2.5	0.4	0.7	0.2	1.2	·
			1978	5.6	0.2	2.3	2.0	0.6	0.5
			1981	1.5	0.2	0.3	0.3	0.7	0.0
11.	G01-G12	Instruments (measuring, testing, optics, photography, holography, horology, controlling, computing, checking devices, information storage, etc.)	1975	4.3	0.5	1.7	0.3	1.8	·
			1978	4.6	0.3	2.1	0.8	1.0	0.4
			1981	1.9	0.1	0.3	0.4	0.8	0.3
12.	G21	Nucleonics (nuclear physics, nuclear engineering)	1975	3.4	0.3	0.8	0.0	2.3	·
			1978	2.5	0.0	0.1	0.1	2.4	0.0
			1981	1.2	0.0	0.1	0.0	1.1	0.0
13.	H01,H02 H05	Generation, conversion or distribution of energy	1975	3.8	0.3	0.8	0.4	2.3	·
			1978	5.9	0.2	2.5	1.1	1.9	0.2
			1981	2.3	0.1	0.2	0.4	1.3	0.3
14.	H03-H04	Basic electronic circuitry and electric communication technique	1975	2.5	0.3	1.3	0.3	0.6	·
			1978	4.1	0.3	1.8	0.8	1.1	0.1
			1981	1.1	0.1	0.2	0.4	0.4	0.0

Source: Own computations based on WIPO statistics

Table 8.9. CMEA Comparative Advantages in Patenting Abroad
(coefficient of comparative advantages)

LPC Class Code	Content	Year	CMEA Total	Czecho-slovakia	GDR	Hungary	USSR	Po-land
1. B01-B09	Separating and mixing	1975	0.8	1.0	0.6	0.6	1.1	.
		1978	2.5	0.5	6.8	2.7	1.4	5.0
		1981	1.1	0.8	0.8	2.1	1.2	3.5
2. B21-B32	Shaping	1975	1.3	0.8	0.8	0.2	2.5	.
		1978	1.9	0.6	4.7	2.9	1.2	2.5
		1981	0.7	0.4	0.9	0.1	0.9	2.5
3. B41-B44	Printing	1975	0.8	3.1	0.9	0.0	0.3	.
		1978	1.0	0.6	4.3	0.7	0.4	0.0
		1981	0.3	0.8	1.1	0.	0.1	0.0
4. B60-B68	Transporting	1975	0.7	0.6	1.1	0.2	0.7	.
		1978	1.6	0.8	4.9	2.7	0.6	0.0
		1981	0.4	0.4	0.4	0.3	0.3	2.0
5. C01-C14	Chemistry	1975	0.7	0.6	0.6	1.5	0.5	.
		1978	1.9	0.8	5.5	3.4	0.7	1.0
		1981	0.8	0.4	0.4	1.7	0.5	1.0
6. C21-C30	Metallurgy	1975	0.9	0.2	0.3	0.6	1.8	.
		1978	2.3	0.8	6.8	3.0	1.2	2.5
		1981	0.8	1.9	2.1	0.4	0.5	0.0

No.	Code	Category	Year						
7.	D01–D07	Textiles	1975	1.2	3.5	1.5	0.2	0.7	.
			1978	1.8	2.2	5.1	1.8	1.0	1.5
			1981	0.3	0.0	0.4	0.4	0.1	1.0
8.	E01–E06	Building	1975	0.6	0.2	0.8	0.5	0.8	.
			1978	2.8	0.0	3.4	4.5	2.0	10.0
			1981	0.3	0.0	0.4	0.4	0.1	1.0
9.	E21	Mining	1975	1.4	0.0	0.0	1.9	2.8	.
			1978	2.8	0.0	3.4	4.5	2.0	10.0
			1981	1.1	0.0	0.0	1.8	0.7	6.0
10.	F01–F17	Engines, pumps and engineering in general	1975	0.6	0.8	0.6	0.3	0.8	.
			1978	1.8	0.6	4.9	3.6	0.4	2.5
			1981	0.5	0.8	0.6	0.4	0.5	0.0
11.	G01–G12	Instruments	1975	1.1	1.0	1.4	0.5	1.2	.
			1978	1.4	0.8	4.5	1.4	0.7	2.0
			1981	0.4	0.4	0.6	0.5	0.6	0.5
12.	G21	Nucleonics	1975	0.0	0.6	0.7	0.0	1.5	.
			1978	0.0	0.0	0.2	0.2	1.7	0.0
			1981	0.4	0.0	0.2	0.0	0.8	0.0
13.	H01,H02, H05	Generation, conversion or distribution of energy	1975	1.0	0.6	0.7	0.6	1.5	.
			1978	1.8	0.6	5.3	2.0	1.3	1.0
			1981	0.7	0.4	0.4	0.5	1.0	1.5
14.	H03–H04	Basic electronic circuitry and electric communication technique	1975	0.6	0.6	1.1	0.5	0.4	.
			1978	1.3	0.8	3.8	1.4	0.8	0.5
			1981	0.3	0.4	0.4	0.5	0.3	0.5

Source: Own computations based on WIPO statistics

question. Higher than average shares emerge in the areas of industrial processing (shaping, separating and mixing), metallurgy, textiles and mining.

Fourth, there are some important differences in the patenting power of individual CMEA countries. The last two observations are well supported by statistics on the comparative advantages of CMEA countries in patenting abroad (see Table 8.9.). The statistics are based on the calculation of the coefficient of comparative advantages, defined as:

$$C_a = \frac{A_i}{B_i} : \frac{A}{B}$$

A_i number of patents granted abroad in class "i" to nationals of A

B_i number of patents granted abroad worldwide in class "i"

A   number of patents granted abroad to nationals of A

B   number of patents granted abroad worldwide

A country enjoys comparative advantages if the said coefficient is greater than one and is subject to disadvantages if it is smaller than one.

The data contained in Table 8.9. indicate that areas of comparative advantages are similar in most of the CMEA countries. Thus, one could hypothesize that the region as a whole represents a common pattern of technological specialization. This observation, if valid, has far reaching consequences. It means that there is little chance for regional technological self-sufficiency, as the weak and strong points of the individual CMEA member countries are largely similar. There are, of course, some exceptions. The GDR, for example, is apparently the only CMEA country which is relatively strong in printing, transporting and in basic electronic circuitry. Hungary, on the other hand, is a clear leader in chemistry and the USSR in nucleonics. The areas of "collective advantages", on the other hand, include primarily industrial processes (separating, mixing and shaping), textiles, mining and energy generation and distribution. These are basically old traditional technologies, supporting mainly the sunset industries of primary industrialization. This means that in all areas of modern technology, CMEA countries have to rely, at least on a medium-term basis, on external sources of technological supply.

It is perhaps worthwhile noting that in the period under examination, there were no special traces of a restructuring process of the patenting power of the said countries, though some positive changes at the end of the 1970s could be detected. These were apparently reversed during the following years.

In addition to measuring areas of comparative advantages in patenting, it seems useful to identify the areas of apparent patenting specialization of the countries. This could be achieved by comparing the national structure of patenting with international standards. The more the patenting structure of a given country differs from the world standards, the more it pursues different directions of technological advances. In view of the lack of proper statistics for all CMEA countries, data for Poland were taken as a proxy for the region as a whole. In order to eliminate the impact of differences in national patenting systems, figures for foreign patenting were used. These have, however, two major deficiencies. First, the patenting structure may be significantly influenced by the intensity of the patent protection (in how many countries a given invention has been patented). Second, only part of the national inventions are normally patented abroad and, in the case of Poland, this is at best 10%. Of course, one can argue that these are normally the most important inventions, at least in the case of the socialist countries, due to difficulties in securing the necessary hard currency financing for patent purposes. Still, however, the sample in question may be called modest at best. Thus, the final conclusions must be formulated with great care. The results of the relevant calculations are shown in Table 8.10.

The picture which emerges, even considering the reservations stated above, is quite instructive.

It turns out that Poland tends to specialize in traditional industrial technologies and equipment (physical and chemical processes, crushing, pulverising, separating, centrifugal apparatus, mechanical metal-working, casting, machine tools, etc.), in inorganic chemistry and fertilizers, and in metallurgy and instruments. Of the four areas of technology mentioned, only the last one, instruments (including measuring, testing, optics, photography), is a representative of the contemporary progressive direction of technological change. The remaining areas are clearly in retreat. This could mean that the directions of technological change in the CMEA countries, as far as potential technology is concerned, are rather outmoded and indicate a technological gap vis-à-vis the developed world. This observation is confirmed by the areas of apparent technological retardation, which include (see Table 8.10.)

Table 8.10. Patenting Specialization of Poland (according to IPC classes)

IPC Class Code	Content	Foreign Patents Granted to Polish Nationals (% of total)		Foreign Patenting World-wide (% of total)		Coefficient of Polish Specialization	
		1978	1981	1978	1981	1978 (3:5)	1981 (4:6)
1	2	3	4	5	6	7	8
1. B01-B09	Mixing and splitting	17.1	13.1	3.0	3.6	5.7	3.6
2. B21-B23	Shaping I	10.0	6.3	2.8	2.8	3.6	2.3
3. B24-B32	Shaping II	4.3	1.8	2.6	3.6	1.7	0.5
4. B41-B44	Printing	0.0	0.0	1.0	1.0	0.0	0.0
5. B60-B64	Transporting I	0.5	1.8	2.7	3.0	0.2	0.6
6. B65-B68	Transporting II	1.1	13.7	3.2	5.0	0.3	2.7
7. C01-C05	Inorganic chemistry	8.4	8.9	2.7	2.5	3.1	3.6
8. C07+A01N	Organic chemistry	3.3	3.9	8.9	9.0	0.4	0.4
9. C08	Macromolecular compounds	4.6	1.8	3.5	3.3	1.3	0.5
10. C09-C11	Petrochemicals	0.0	0.8	2.2	2.3	0.0	0.3
11. C12-C14	Pharmaceuticals	0.0	0.0	0.5	0.9	0.0	0.0
12. C21-C23 and C25-C30	Metallurgy	3.0	5.7	2.1	2.2	1.4	2.5
13. D01-D07	Textiles	3.3	0.9	2.1	1.8	1.6	0.5
14. E01-E06	Building	1.6	2.7	2.7	2.0	0.6	1.4
15. E21	Mining	0.5	0.1	0.5	0.8	1.0	0.2
16. F01-F04 and F15	Engines, pumps, hydraulics	6.5	0.6	2.2	2.6	3.0	0.2
17. G01-G03	Instruments	10.0	10.1	6.5	6.0	2.2	1.7
18. G21	Nucleonics	0.0	0.0	0.04	0.4	0.0	0.0
19. H01,H02, H05	Generation, convertion or distribution of energy	0.8	11.0	5.4	6.2	0.1	1.8
20. H03-H04	Basic electronic circuitry and electric communication technique	1.6	0.6	2.7	3.1	0.6	0.2

Source: Own computations based on WIPO and Polish patent statistics

printing, transporting, organic chemistry, pharmaceuticals, petro-chemicals, nucleonics, basic electronic circuitry and telecommuni-cation.

5. Evolution of the CMEA Position in Applied Technology

As argued before, the level of applied technology, i.e. technology actually in use, may be measured with two sets of indicators: one stemming from the production sphere and another related to for-eign trade statistics.

Production-related indicators cause severe difficulties in cross-country comparisons, mainly due to the distortional impact of the price variables and, additionally, in the case of CMEA countries, due to the non-convertibility of their national currencies and the resulting lack of an economically meaningful rate of exchange. Therefore, the relevant indicators must be selected with great care, taking into account the aforementioned difficulties.

5.1. Production-related Indicators

A standard measure frequently applied to assess the changes in the overall level of technology used, is the productivity variable. It reflects the principal long-term target of the technological change, i.e. an improvement in the use of the available resources. It is common practice to measure productivity changes in terms of labor productivity due to both statistical (technical) and sub-stantial reasons (3). On the technical side, one could mention the relative abundance of pertinent data and the simplicity of the re-quired calculations. On the substantial side, it should be noted that the labor force constitutes one of the most important eco-nomic resources.

Now, if we look at the CMEA performance in labor productiv-ity over the last two decades, we can conclude that, on an aver-age, it has been reasonably successful and that the CMEA coun-tries have managed to achieve a rate of growth considerably higher than the OECD countries (see Table 8.11.).

Table 8.11. Long-term Trends in Labor Productivity (Average annual growth rate)

Regions and Countries	1961-73	1974-79	1980-83	1984	1985
1. Western Europe[a]	4.2	2.2	1.2	2.4[f]	2.2[f]
2. USA[a]	2.0	0.1	0.2	2.9	0.5
3. European CMEA[b] (4+5)	5.1[c]	3.3[d]	2.8[e]	4.7[g]	3.2[g]
4. Eastern Europe[b]	6.1[c]	3.4[d]	2.1[e]	4.3[g]	3.2[g]
5. USSR[b]	4.6[c]	3.3[d]	3.1[e]	2.8	3.4
6. Bulgaria		6.2	3.0	5.1	2.0
7. Czechoslovakia		3.3	0.6	2.8	2.4
8. GDR		3.6	3.7	5.2	4.5
9. Hungary		3.7	3.1	3.6	1.0
10. Poland		0.9	-3.2	5.9	3.2
11 Rumania		7.2	2.7	7.5	5.9

a) Output per person employed in the economy as a whole
b) Output per person employed in the material sphere
c) 1961-75
d) 1976-80
e) 1981-84
f) Unweighted average of the 13 West European countries
g) Unweighted average

Source: Calculated from Economic Survey of Europe in 1985-1986, UN, New York 1986, pp. 4-8; p. 120; H. Wienert, J. Slater - East-West technology transfer. The trade and economic aspect, OECD 1986, pp. 405-411

The growth rate of their labor productivity was most spectacular from 1961-1975, decreasing substantially in the following period. Interestingly enough, the changes in the growth rate of labor productivity in CMEA countries showed much the same trends as observed in OECD countries.

As far as individual CMEA countries are concerned, the best results were recorded by Bulgaria, Rumania and the GDR, while Poland and Czechoslovakia had the worst results.

Notwithstanding the relatively high growth rate of labor productivity in CMEA countries over the last two decades, its absolute level still remained far behind that recorded in OECD countries. According to available estimates, the absolute level of labor productivity in the CMEA area, as a whole, was 3.4 times lower than the average for OECD in 1960, and was still 2.6 times lower in 1979 (4).

Apart from the labor productivity, a reasonably unbiased picture characterizing both the overall level of applied technology and its changes over time can be achieved with the help of figures related to the material and energy intensity of the national output. Calculated in physical units, the figures avoid the pitfall of price-bias. At the same time, due to their nature, they seem to capture different aspects of the level of technology applied in different production branches of the national economy. A major line of argumentation pursued in this respect is that the lower the material and energy intensity level of the national income, the higher the level of technology actually in use.

As far as the material intensity of the national output is concerned, one can measure it with the help of the steel intensity, i.e. the apparent steel consumption per unit of NMP produced. The CMEA countries, as a whole, managed to decrease their absolute steel intensity in 1970-1984 by about 30%. More successful results were achieved by East European countries (a reduction of 32%), while the Soviet Union had less success (25%). Still, however, the absolute level of the steel intensity was high (see Table 8.12.).

As indicated in Table 8.12., in 1980, the said countries consumed, on an average, 2.6 times more steel per $ 1000 of their national income, than the developed countries of Western Europe. The same ratio for such countries as Poland, the USSR and Czechoslovakia was, however, much higher, thus indicating the size of the technological gap still existing.

Of course, one should be aware of the fact that the high steel consumption per unit of national income reflects not only the level of technologies in use, in a broad sense (both products and processes), but also the industrial structure of the country (so-called product mix). Thus, countries pursuing the development of steel-consuming industries would inevitably arrive at a higher coefficient of steel consumption. For example, a relatively high consumption of steel can be found in Czechoslovakia and the USSR from the CMEA and Italy and FRG among the OECD countries.

Table 8.12. Steel Consumption per Unit of National Income (Eastern and Western Europe) in 1980 (a)

CMEA	Steel Intensity of National Income	West	Steel Intensity of National Income
Bulgaria	87	Austria	39
Czechoslovakia	132	Belgium	36
Hungary	88	Denmark	30
GDR	88	FRG	52
Poland	135	France	42
USSR	135	Netherlands	31
CMEA Total (weighted average)	111	Italy	79
		UK	38
		Western Europe Total (weighted average)	43

(a) In kilograms of steel per $ 1000 of national income

Source: J. Monkiewicz - Technika polska w kontekscie miedzynarodowym, Zagadnienia Naukoznawstwa, 3(83), 1985, p. 374

Much the same picture can be observed with respect to the energy-intensity of the national income of the countries concerned (5). Although this declined in most of the CMEA countries from 1976-1984, it was still much higher than in Western countries (see Table 8.13.).

As indicated in Table 8.13., the energy-intensity of the national income produced in the CMEA countries was, on an average, over two times higher than in nine EEC countries for the period 1970-1980. Interestingly enough, the differences increased slowly but surely throughout the entire period, with a marked deterioration of the CMEA position in 1979-1980.

Table 8.13. Energy Consumption per Unit of National Income (a)

Country	1970	1974	1979	1980
Bulgaria	2064	1924	1783	1788
Czechoslovakia	2505	2240	2111	2026
GDR	2582	2256	2110	2047
Poland	2335	1825	1968	2118
Hungary	1796	1642	1573	1459
USSR	2374	2324	2253	2239
CMEA Total	2369	2240	2179	2171
Belgium	1221	1142	1001	920
Denmark	1008	821	847	783
France	984	923	870	852
Netherlands	1092	1137	1107	1126
Ireland	1618	1375	1475	1394
Luxemburg	3544	3390	2723	2726
FRG	986	939	880	818
U.K.	1748	1572	1438	1376
Italy	1193	1192	1242	1150
9 EEC Total	1171	1099	1242	1150

(a) In gram of coal equivalent per $ 1 of national income

Source: J. Danielewski - Energetyczne problemy RWPG, IKiC, Warszawa 1983, p. 177

In light of the rather modest results recorded in CMEA countries with respect to the shifts in the overall level of applied technology over the last 20 years, quite a different picture is offered in terms of the CMEA position in some key areas of current technological progress. This is especially true of the production and use of industrial robots and flexible manufacturing systems.

According to recent ECE estimates, CMEA countries, as a whole, possessed around 21% of the total worldwide programmable robot population in 1983. Most of these were located in the Soviet Union (6). It was estimated that by 1985, the programmable robot population in CMEA countries would increase to ca. 21,000 or 18.6% of the world total (see Table 8.14.).

Table 8.14. Estimated Growth of World Robot Population (units)

Country/Region	1981	1982	1983[a]	1985[a]
Japan	8500	12000	16500	35000
US	7000	9000	8500	28000
Western Europe	6000	9000	13000	25000
CMEA	6000	8000	10500	21000
Others	1500	2000	1500	4000
Total	24000	40000	50000	113000

(a) Estimate

Source: ECE/ENG.AUT/15, pp. 44-45

The recently announced robotization plans in the USSR, the GDR, Czechoslovakia, Poland, Bulgaria and Hungary indicate not only a growing awareness among the CMEA policy-makers as to the importance of this new technology, but may also be taken as proof of the political will to increase the number of robots at a faster pace.

As far as flexible manufacturing systems (FMS) are concerned, the CMEA countries are apparently among the pacesetters of new technology (7). According to a recent ECE study, more than a quarter of the world's total FMS population is located in the CMEA region (see Table 8.15).

Table 8.15. World population of Flexible Manufacturing Systems (FMS) in 1984/85

Country	Units	% total
Bulgaria	7	1.9
Czechoslovakia	12	3.3
Hungary	7	1.9
GDR	10-20	2.8-5.6
USSR	60	16.7
Sub-total	96-106	26.0-28.8
France	17	4.7
Netherlands	25	7.0
Sweden	11	3.1
U.K.	3 (35)	0.8
FRG	25-35	7.0-9.7
Italy	25	7.0
Japan	100	27.9
US	47	13.1
Sub-total	253-263	70.6-73.3
Grand Total	349-369 (35)	100

Note: The figures in parentheses indicate the number of installations under construction

Source: ECE/ENG.AUT/22, p. 25

The Soviet Union is unquestionable regional leader which installed its first FMS already back in 1968. After that, 13 FMS's were installed from 1971-80, 10 during 1981 and 30 in the period 1982-1983 (8). It is followed first by the GDR, which plans to build more than 100 new FMSs from 1986-1990 (9), and then by Czechoslovakia.

Definitely less spectacular results were recorded by the CMEA countries in the computer industry, particularly when measured in international terms. However, the achievements of the CMEA countries in this area still cannot be denied. They have managed to survive hard embargo measures applied by the US and other OECD countries and have succeeded in building up an indigenous computer industry, practically self-sufficient for regional needs. In contrast to most OECD countries, they have managed to avoid the domination of the US producers, though no doubt with some costs both with respect to the necessary effort and the technical level. According to the last available estimates, CMEA countries controlled ca. 6.8% of the worldwide market for mainframe computers in 1984, as opposed to 6.1% in the case of Japanese and 4.8% in the case of European producers. The remaining 82.3% of the market share belonged to the US manufacturers, 58.4% held by IBM alone (10). With regard to the CMEA regional self-sufficiency in computer equipment, it may be interesting to note that the West European computer market is controlled by local producers only to 14.6%, with the remaining part of the market in the hands of the US manufacturers, of which 59.8% is controlled by IBM alone (11).

5.2. Trade-related Indicators

Measuring the dynamics in the levels of applied technology in CMEA countries using trade-related indicators has some clear advantages over using production indicators. The most important of these is the fact that as long as data for extra-CMEA trade are applied, the distortionary impact of national price systems and of artificial exchange rates are eliminated. The relevant procedures may be based on a number of more or less refined statistical methods, two of which seem to be especially useful: one is based on the analysis of export structures and the other on the analysis of export prices. The first assumes the existence of a direct relationship between the qualitative structure of exports and the level of technology in use. It is argued that the larger the share of technology-intensive products in the country's exports, the higher its technology level, viewed from an international perspective (12).

The second, on the other hand assumes, that the higher the export prices per unit (per piece or better per weight), the higher the level of technology actually in use.

These two approaches clearly have their theoretical justification in the neo-technology theories of international trade as formulated by Posner, Hufbauer, Vernon and their followers.

The first more refined analysis of the qualitative structure of East-West trade was provided in 1978 by the Economic Commission for Europe. In this study, covering the period 1965-1977, all products appearing in East-West trade were divided into four distinct groups:

- natural resource intensive products (A),
- capital intensive products (B1),
- labor intensive products (B2), and
- technologically advanced products (C) (13).

The last group, which is the major area of our concern, included: organic and inorganic chemicals, medicinal products, plastics and synthetic materials, miscellaneous chemicals, power machinery, agricultural machinery, office machines, metalworking machines, special industrial machines, electric power distributing machinery, instruments and related apparatus, photographic supplies, watches, clocks, ordnance and ammunition.

The study indicated that in the period 1965-1977, CMEA countries, as a whole, increased the share of technologically advanced products (TAP) in their western exports, from an average of 9% from 1965-1969, to an average of 12% from 1973-1977. The respective figures for Soviet exports was 5% and 7% (14). It is perhaps worth noting that at the same time, the share of TAP in the CMEA western imports was much higher, reaching 48% from 1965-1969 and 46% from 1973-1977. The respective figures for the Soviet Union were 49% and 44%.

Much the same results were obtained in another study on the share of TAP in Polish western exports, covering the period 1971-1980 (15). This study also revealed that the share of TAP in Polish exports to developing countries remained fairly constant from 1971-80, amounting to around 29% throughout the period. On the other hand, the share of TAP in Polish imports from developing countries dropped from 4% in 1971 to 2% in 1980 (16).

The trends detected in the aforementioned studies are largely confirmed by recent UNCTAD findings in this area. Using a much narrower definition of technology intensive products (called R+D

Table 8.16. OECD Imports from all Countries Manufacturing by R and D Intensity

Country of Origin	(in million of US $)					
	1970	1975	1980	1982	1984	1985
High R+D						
World	18271	44868	120772	118414	147227	158196
Social. Countries of East Europe	111	436	1423	945	838	863
from which USSR	20	131	808	465	341	363
Medium R+D						
World	55696	129576	320368	290297	335326	371717
Social. Countries of East Europe	677	1991	4607	3694	3889	4083
from which USSR	162	605	1604	1277	1389	1347
Low R+D						
World	72572	174640	383086	345234	371300	380955
Social. Countries of East Europe	2655	7747	1796	16804	17321	17053
from which USSR	900	3240	8043	9121	8369	8408
Total manufactures from CMEA, of which	3443	10174	23994	21443	22048	21999
- high R+D (%)	3.2	4.3	5.9	4.4	3.8	3.9
- medium R+D (%)	19.7	19.6	19.2	17.2	17.6	18.6
- low R+D (%)	77.2	76.1	74.9	78.4	78.6	77.5
Total manufactures from the Soviet Union, of which	1082	3976	10455	10863	10099	10118
- high R+D (%)	1.8	3.3	7.7	4.3	3.4	3.6
- medium R+D (%)	15.0	15.2	15.3	11.8	13.8	13.3
- low R+D (%)	83.2	81.2	77.5	83.9	82.8	83.1

Source: Calculated from data supplied by UNCTAD Secretariat

intensive products), UNCTAD indicates that the share of TAP in CMEA manufacturing exports to OECD countries grew relatively fast throughout the period 1971-1980. Thereafter, however, it fell significantly, but still exceeded the one registered at the outset of the 1970s. Particularly dramatic changes took place with Soviet TAP exports (see Table 8.16.).

Parallely to the changing share of TAP in CMEA exports to OECD, there was a change in the CMEA market share in OECD in this product category. It climbed from 0.61% in 1970 to 1.18% in 1980 and then declined to 0.55% by 1985.

The three studies in question were based on a broad definition of technology intensive products and did not provide the necessary information for an analysis of the relative export performance of individual CMEA countries. Therefore, it seems useful to supplement these with an analysis of TAP exports in a narrow sense and extend it to cover the export performance of individual CMEA countries. The most popular concept of TAP's in a narrow sense is the one elaborated by the US Administration and more specifically, by the US Department of Commerce (17). The list of products includes primarily items from the SITC group 7 (machinery and transport equipment) and some from SITC 8, and excludes, in contrast to the ECE concept, all chemicals and intermediates.

Taking this definition as a point of departure, the CMEA TAP exports to five leading Western countries from 1971-80 were analyzed (see Table 8.17). According to the findings, the total export volume of the said products increased nearly 5 times from 1971-80, slightly more than the total export of machinery and transport equipment to the OECD area from 1971-1980 (which increased 4.4 times). It is worth pointing out that there were substantial differences in the export performance of individual CMEA countries.

Czechoslovakia, the leading exporter of TAP in 1971, and the Soviet Union, second on the list in 1971, recorded substantial losses in their export shares in 1980, whereas the largest increases were recorded by Rumania, Poland and Hungary. This might indicate a changing technological hierarchy within the CMEA region from 1971-1980, an observation which will be qualified in the subsequent discussion.

As far as the product-mix of the TAP exports is concerned, the dominant item throughout the entire period was machine tools for metals. The second largest group (in value terms) constituted rolling and ball bearings. Other products, like pumps and centrifuges, electrical instruments, transistors and photocells, played a much smaller role in terms of value. Thus, on the whole, the said exports were limited foremostly to traditional TAP products, with a poor representation of the new expanding industries.

Table 8.17. Export of TAP from CMEA to 5 Leading Western Economies (USA, FRG, Italy, France and U.K.)

Country	1971 $ 1000	% of total	1980 $ 1000	% of total	1980:1971
Bulgaria	2861	3.5	18381	4.6	6.4
Czechoslovakia	26057	31.8	76904	19.4	3.0
GDR	13594	16.6	47248	11.9	3.5
Hungary	6930	8.5	49143	12.4	7.1
Poland	11247	13.7	100628	25.3	8.9
Rumania	4362	5.3	58449	14.7	13.4
USSR	16761	20.5	46260	11.7	2.8
Total	81812	100.0	397013	100.0	4.9

Source: J. Monkiewicz, J. Maciejewicz - Technology export from the socialist countries, Westview Press 1986, p. 57

Let us now turn to the issue of measuring shifts in the technology level via the level and changes of export unit prices, i.e. prices per weight of the exported products. The method in question does not provide indications on the absolute level of applied technology, but rather on its shifts, in comparison to other countries or regions. The analysis which follows was based on EEC foreign trade statistics and covered all 12 sections (out of a total 21), which represent the products of manufacturing industries. EEC countries have a major share in East-West trade so the results obtained serve as a good proxy for the total East-West trade. Both import and export prices were calculated and it was assumed that changes in the ratio of export to import prices measure the changing technological distance between the CMEA and developed capitalist countries. The results obtained are quite instructive (see Table 8.18.).

Table 8.18. Changes of Export/Import Unit Price Ratio in the CMEA Trade with EEC Countries

Sections of EEC Foreign Trade Classification	1976	1980	1984	1985	1985/ 1976
1. Section V (Products of mineral industry)	0.525	0.832	1.458	1.480	2.8
2. Section VI (Products of chemical industry)	0.329	0.261	0.195	0.173	0.5
3. Section VII (Plastics, etc.)	0.468	0.434	0.386	0.360	0.8
4. Section VIII (Skins, hides and products)	1.713	2.224	1.820	1.272	0.7
5. Section IX (Wood and wooden products)	0.329	0.656	0.027	0.023	0.7
6. Section X (Paper and derivatives)	0.137	0.257	0.280	0.261	1.9
7. Section XI (Textiles)	0.540	0.742	0.718	0.739	1.4
8. Section XIII (Ceramics, glass, lime)	0.255	0.272	0.305	0.325	1.3
9. Section XV (Metals)	0.588	0.652	0.721	0.766	1.3
10. Section XV (Machinery and equipment)	0.265	0.279	0.208	0.204	0.8
11. Section XVII (Transport equipment)	0.665	0.620	0.440	0.384	0.6
12. Section XVII (Measuring and controlling instr.)	0.272	0.187	0.192	0.158	0.6

Source: Calculated from analytical tables of external trade, NIMEX, Eurostat 1976-1986. Calculations were performed by O. Bagniewski from Hamburg University

Table 8.19. Changes of Export/Import Unit Price Ratio in Brazilian and South Korean Trade with EEC Countries

Content	Brazil						South Korea			
	1976	1980	1984	1985	1986	1985/1976	1980	1984	1985	1985/1976
1. Section V	0.165	0.125	1.258	0.138	0.8	2.612	2.347	5.716	2.021	0.8
2. Section VI	3.785	2.471	1.660	0.912	0.2	3.089	3.397	3.073	2.452	0.8
3. Section VII	6.684	1.236	-	0.586	0.1	1.370	1.220	1.362	1.246	0.9
4. Section VIII	1.295	1.172	0.757	1.336	1.0	3.058	1.317	0.971	0.938	0.3
5. Section IX	0.429	0.345	0.481	0.230	0.5	-	0.732	1.502	0.552	.
6. Section X	0.192	0.213	-	0.746	3.9	0.451	0.308	0.425	0.520	1.2
7. Section XI	0.600	0.579	1.408	0.527	0.9	1.598	1.044	2.086	1.270	0.8
8. Section XIII	2.554	0.692	1.339	0.346	0.1	0.523	1.618	2.430	2.409	4.6
9. Section XV	0.350	0.386	0.650	0.331	0.9	0.237	0.806	4.231	1.653	7.0
10. Section XVII	1.341	0.331	1.482	1.129	0.8	0.269	0.255	3.501	4.221	15.7
11. Section XVIII	0.389	0.577	9.063	0.447	1.1	0.303	0.244	0.261	0.217	0.7

Source: As in Table 8.18

First, it is worth pointing out, that from the 12 large product groups (sections), CMEA countries, as a whole, recorded constantly higher export than import prices in only one (skins, hides and products thereof), and since the mid 1980s it has been supplemented by another product group, i.e. products of the mineral industry.

Second, apparently the worst results in the export/import ratio were recorded in more high-tech industries: chemicals, plastics, machinery and equipment, measuring and controlling instruments as well as transport equipment.

Third, in all high-tech sections there was a deterioration in export/import unit prices over the period 1976-1985. All of these factors imply, according to our assumptions, that the international technological position of CMEA, vis-à-vis developed capitalist countries, was deteriorating.

It may be interesting to compare these developments with the performance of Brazil and South Korea on the same market in the same product groups (see Table 8.19.).

Without going into unnecessary detail, it is worth noting that the two countries, taken as the leading representatives of the NIC family, have on the whole recorded much better results than the CMEA countries together. This is well proved by the fact that they attain an export/import unit price ratio higher than 1 more frequently. In the case of Brazil, such a positive ratio was recorded regularly in 3 sections, two of which belong clearly to modern industries. In the case of South Korea, export unit prices were higher than those of import in six sections, including chemicals, plastics and machinery and equipment. Additionally, in most of the remaining sections, the two countries achieved better export/import unit price ratios than the CMEA total.

What is true for both CMEA and the two NICs, however, is that the said ratio has on the whole deteriorated over the period concerned. This could mean that the said countries have also lost their initial technological impetus and increased the technological distance vis-à-vis the EEC countries.

The two countries had not only a better export/import unit price ratio, but also much better export unit prices than the CMEA total. Ranking the prices received per unit for their respective exports and assigning different weights for subsequent rankings, we can identify a global picture in the said area (see Table 8.20.).

It may be somewhat surprising, but the results of the relevant calculations point out that Brazil and South Korea have been achieving much better export prices than the CMEA total already since 1970, and that the observed proportions have remained basically unaltered since then.

Table 8.20. Global Assessment of the Relative Technological Level of CMEA's, Brazil's and South Korea's Exports to EEC, in 12 Sections

	1970	1976	1980	1984	1985
CMEA	8	10	8	6	6
Brazil	20	20	20	22	20
South Korea	32	30	32	32	34

Note: Weighs 3 points for rank 1   maximum gain possible =
          2 points for rank 2   36 points
          0 points for rank 3

Source: As in Table 8.18.

We can use the same type of reasoning for the analysis of the intra-CMEA shifts in the technological position of the individual member countries. Let us begin with export unit prices. Again, ranking the countries according to the level of the prices received and assigning relevant weights, we can arrive at a global assessment of the changes in the technological level of the relevant exports. The results of the calculations are provided in Table 8.21.

The picture emerging is quite instructive. First, it seems that in spite of the different technological policies pursued by individual CMEA countries, the relative technological level of their exports remained basically unaltered with some ups (Poland during 1981-84) and downs (Czechoslovakia during 1980-1984). The only clear loser is Bulgaria while the unquestionable leader over the whole period was Hungary. An astonishingly high position was held by Rumania, which in fact managed to slightly improve its technological performance at the end of the 1970s and the beginning of 1980s. Much the same picture appears when export/ import unit price ratios are taken into account.

Table 8.21. Global Assessment of the Relative Technological Level of the CMEA Countries' Exports to EEC, in 12 Sections

Country	1970	1976	1980	1981	1982	1984	1985
Bulgaria	32	24	20	30	20	22	21
Czechoslovakia	18	25	16	10	10	13	22
Hungary	41	40	45	44	44	39	40
Poland	18	18	19	21	26	25	18
Rumania	23	25	32	27	32	33	31

Note: Rank 1 - 5 points   maximum gain possible = 60 points
      Rank 2 - 3 points
      Rank 3 - 2 points
      Rank 4 - 1 point
      Rank 5 - 0 points

Source: As in Table 8.18

6. Conclusions

Assessing the dynamics of CMEA's technological position, in international terms, over the last two decades, we can note first of all that the picture emerging from the statistical analysis is extremely differentiated. It is very positive as far as the scientific and technical infrastructure is concerned. The CMEA countries managed to build up and maintain vast R+D sectors, with R+D employment much higher than in the industrialized western economies. The extent of their R+D effort also clearly exceeded that observed in the rest of the world throughout the 1970s and 1980s.

The same picture is, however, less positive if we start to look at the primary results of the R+D effort, i.e. the patenting output. In quantitative terms, it is still impressive but in qualitative terms, it is much poorer. The region as a whole has neither managed to attract many foreign patents nor has it managed to improve its position in foreign patenting. Both its foreign patenting attractiveness and foreign patenting aggressiveness are much below the standards observed in the developed countries. Even if this situation could be partly attributed to policy and institutional-related variables, a substantial part could still be related to the lack of meaningful, commercially viable inventive output. The said picture is even darker if the sectoral structure of the patenting activity is taken into account. Available indicators point out that the CMEA countries as a whole tend to exhibit their comparative advantages over the rest of the world in old, traditional fields of technology, with a clear under-representation of the most modern, currently leading industries. This observation has far reaching consequences as it indicates that the local technological resources (potential technology) will be insufficient for the future development of modern industries and thus, the possible restructuring process would have to be based to a considerable extent on imported inputs.

CMEA's technological advances in applied technology seem to be even less convincing. It is true that over the last two decades the countries in question have managed to improve their level of applied technology and reduce, to some extent, the distance vis-à-vis the developed countries, but the rate of this improvement has been far too slow to exert any dramatic impact on their international economic standing. Moreover, the rate of this improvement seems to fall behind that observed in the most dynamic segment of the world economy, i.e. the newly industrialized countries. In effect, the latter have managed to replace the CMEA countries in some products and foreign markets and thus have grasped more sophisticated parts of the said markets. The overall picture, however, is not that simple. Having serious problems in the dynamic improvement of their overall technological level, the CMEA countries have managed to defend or to build up a relatively good international position in some key areas, such as computers, robots and flexible manufacturing systems.

Their diffusion, however, throughout economic activities is far from satisfactory and hence, their impact on the overall economic performance is limited.

References and Footnotes

1   R.W. Campbell - Soviet R+D statistics 1975-1982, Indiana University 1982, mimeo, p. 17
2   Report of the CMEA Secretariat, 1982, p. 48
3   See Economic Survey of Europe in 1985-1986, ECE, UN, New York 1986, pp. 4-5
4   GUS, Zaklad Badan Statystyczno-Ekonomicznych, Zeszyt 126, Warszawa, wrzesien 1982
5   Economic Survey of Europe in 1985-1986, ECE, UN, New York 1986, pp. 125-127
6   Production and use of industrial robots, ECE, UN, New York 1985, p. 44
7   Recent trends in flexible manufacturing, ECE, UN, New York 1986, pp. 25-28
8   The FMS Magazine, April 1985
9   The Engineer, March 21, 1985
10 Zur Position der informations- und kommunikationstechnischen Industrie in der Bundesrepublik Deutschland, Nixdorf Computer, Hamburg 1986, mimeo, p. 14
11 Ibid., p. 14
12 See, for example, Report of the President on US Competitiveness, Office of Foreign Economic Research, US Dept of Labor, Sept. 1980, pp. IV5-IV7
13 Economic Bulletin for Europe, Vol. 30, No. 1, UN, New York 1978, pp. 99-101
14 Ibid., p. 100
15 See J. Cieslik, J. Misala - Struktura wymiany towarowej Polski w 1971-80 wedlug intensywnosci czynników produkcji, Warszawa 1982, SGPS, mimeo, p. 12
16 Ibid., p. 12
17 See: Technology and East-West Trade, Congress of the United States, Office of Technology Assessment, Washington DC, 1979

Part IV
Technology Import and the Technology Gap: Major Findings and Conclusions

As indicated in the introduction to this book, the major goal of my analysis was to shed more light on the relationship between technology import and technological backwardness, or more precisely, on the use of technology import as a tool to reduce the technological backwardness of a country. To do this, the recent experiences of the CMEA countries in this area have been taken as the empirical base and testing ground for the subsequent deliberations. In what follows, I shall try to review the major findings and spell out the resulting conclusions.

1. Summary of Major Findings

The point of departure of the analysis performed in this book was the assumption that one <u>cannot properly understand the subsequent handling of foreign technology in CMEA countries and elsewhere without clarifying what factors basically determine national technological performance</u>. In the end, it is precisely the national technological performance, or better the differences in national technological performance, which shapes in the long run the technological leads and lags in the international community. The characteristics of the national technological performance also set, to a considerable extent, the framework for the subsequent technology importation programs.

Particular attention in investigating this question has been given to the role of <u>institutional or systemic factors</u>. This was justified by the widely held belief both in the West and East, that precisely these systemic factors are responsible for CMEA's unsuccessful performance in the production of endogenous technologies and the utilization of foreign technologies.

The analysis performed covered three broadly defined theoretical approaches:

- production- (transaction-) specific approach
- institutional approach
- historical approach

The <u>production-specific approach</u>, still representing the mainstream of relevant theoretical thinking, views national technological performance as a specific process of the production and utilization of technological knowledge, with the relevant actors and stages. This process is supposed to have its own internal logic and is governed by relevant laws and interrelationships and its final effects are to depend on the fulfilment of the necessary requirements, the provision of appropriate factors and on the accomplishment of the correct composition.

The analysis of the production-specific approach revealed, that in spite of its important contribution to our better understanding of national technological performance, it fails to recognize and explain instances in which the same innovative effort brings highly different results, and does not answer the questions of what determines the scope of national innovative effort itself, or how an innovation conducive environment is created, etc. This approach is not, in fact, directly involved in cross-country comparisons in technological performance and, hence, it does not look for national differences in this area, but rather tries to detect and pull together commonalities. This is precisely where the institutional approach comes to the forefront, being more concerned with international differences.

The essence of this approach lies in associating national technological performance with the specific features of the national economic organization. It is claimed, that to understand varying national performances in technological change, systemic variables (system data) must come to the forefront of the analysis.

In assessing the institutional interpretation of various technological performances of different nations, the positive contributions were stressed by bringing to attention the characteristics of the national agents and not only to the transaction itself (production-oriented approach), which shed some new light on the varying resource effectiveness. This helps to clarify and understand the behavior of planned and market systems and, thus, is specially useful for East-West comparisons. At the same time, however, it has some obvious weaknesses that require proper qualification. These include principally the assumption of symmetry (the same elements of different systems are given the same weight, though they may play quite different roles in different systems), the neglect of socio-cultural and developmental variables and the high level of abstractness of the analysis.

The historical approach, on the other hand, draws special attention to the time and place elements in explaining national technological performance. Thus, it places emphasis on the development level, policy measures, sudden disruptions in the accustomed sources of supply, cultural values and social structures. Filling the gap left by previous theoretical schools, the historical approach contributes significantly to the enrichment of the analytical framework. Its weakness, however, is the concentration on ad hoc, time and place factors, which do not provide a sufficient basis for firm generalizations.

The final conclusion resulting from this research was that to understand national technological performance, one needs to consider three interrelated elements: the institutional framework, transactional characteristics and environmental variables of the relevant nations, all taken in the proper proportion. It was concluded that what is needed is an interdisciplinary and ecclectic approach which would make proper use of the propositions contained in all three aforementioned methodological approaches.

The national technological performance, however, is not by any means entirely determined by national variables, but remains under the strong influence of the international system. Therefore, an analysis of the theoretical framework explaining basic relationships among the technological backwardness, importation of technology and the national technological performance was subsequently conducted. It was intended to provide an international setting for understanding nationally organized processes. It was assumed that in order to properly understand the impact of imported technology on the national technological development, a theoretical framework describing the nature of the transaction (technology transfer) and the nature of the mechanism set in motion by technology importation must be clarified.

In doing so, special attention was given to two largely competitive families of theoretical models: technology gap/catching up approach and the dependencia school. Needless to say, the arguments of the authors dealing directly with East-West technological relations (survival hypothesis versus selling the rope hypothesis), were given due consideration.

A principal feature of the technology gap/catching up approach is its generally positive stance with respect to the application of foreign technology to the national technological development, implying its positive impact on national technological change as well as the resulting national economic progress. According to this line of thinking, the opportunities inherent to industrialization vary according to the backwardness of a country. The greater the backlog of technological innovations which a backward country could take over from more advanced nations, the brighter its technological growth prospects.

An important extension of the technology gap/catching up approach is the "resonance" model, which supplements the former with a more detailed analysis of the mechanism through which imported technology influences domestic technological performance.

However, both, the "resonance" and the technology gap models seem to overlook some important factors pertinent to the technology transfer process. The first, is the role of the transfer channels applied. One can argue that the "technological energy" introduced by foreign technology will vary depending on how it is introduced. Compact technological systems (for example, turn-key installations), or the compact nature of technology suppliers (for example TNCs) may significantly alter the final effects of the import operation.

The second is the role of the international economic and political setting or, more generally, the international environment which may have an important influence on which technologies are available and on what terms.

In contrast to the technology gap/catching up theories, which ultimately viewed technological underdevelopment as a kind of economic resource and thus blessed the activities aimed at utilizing foreign developed technologies, the dependency school frequently argued for quite opposite actions. The principal contention of this theory is that the import of foreign technology from developed to backward countries does not serve to decrease technological dependence, but results rather in perpetuating and strengthening it.

According to this theory, the world economy is made up of a number of unequal partners, i.e. dominant and dominated economies resulting finally in the appearance of the domination effect. This effect may come up inter alia due to a sheer difference in the economic potentials of the countries/group of countries concerned, or due to the unequal distribution of some basic resources. In both cases, the dominant economy influences the developments in the dominated countries (consciously or not), thus reducing their freedom of movement and compelling them to certain reactions. Given the lack of motives for the dominant economies to share their resources with dominated systems, or the lack of the latter's ability for absorbing (creating) them, or both, a number of significant consequences arise, the perpetuation of dependency being the most important one.

A separate line of theoretical debate on the relationship between technology import and national technological development is concerned directly with the socialist countries. Two extreme groups of views are identified here. One, which views foreign technology as a life preserver for otherwise clumsy economic systems of the centrally planned economies, principally associated with the proponents of institutional thinking. The second, on the other hand, sees the technology import from the West as a rope of dependency and clearly reflects the concerns of the dependencia thinkers.

The analysis performed pointed out that to properly assess the performance of the CMEA countries as technology importers, a new approach based on the application of the theories of economic power and dominant economy was necessary.

The subsequent analysis allowed the conclusion that East-West technology transfer occurs under the dominant position of the Western economies and results in the domination effect. The nature of this effect is determined by the fundamental incompatibility of political and ideological aims and values of the parties concerned. Thus, East-West technology transfer always was and always will be confronted with a non-conducive international environment, which of course, influences the subsequent absorbtion activities.

Investigating the size and structure of Western technology inflow to the CMEA region in the 1970s, it was concluded that there was a significant increase still, however, it was rather limited by international standards. Between 1970-1975, the value of technology infusion more than doubled, thereafter, however, it dropped in real terms. This may indicate that technology import was not among the top priority objectives of the CMEA policymakers. The importance of Western technology was further reduced by its relatively low level of sophistication. On an average, only 2.5% of all technology inflows were of high R+D intensity and more than 3/4 of the total were low R+D products.

The analysis indicated that throughout the entire period the principal methods of importing Western technology were traditional pure foreign trade transactions and the import of capital goods. Regulated forms of technology infusion (i.e. various types of international cooperation arrangements) remained quantitatively insignificant, though their emergence heralded some new qualitative changes in East-West capital links. It is frequently argued that the movement of technology across borders is inseparable from the transborder movement of investment capital. Logically then, obstacles of foreign investments become obstacles of international technology flows. An analysis of the existing foreign investment ventures of the CMEA countries revealed a very peculiar picture, namely that their technological characteristics (technological intensity) were below that registered for regulated forms of East-West technology flows. Thus, it undermines the observations offered so far in the related literature. In explaining this peculiarity, three hypotheses were formulated: the half-pregnant approach (all or nothing explanation), the asymmetrical cost-benefit configuration (insufficient motivations offered), and the Wild West hypothesis (adventurer, time-period explanation). The first one claims that offering foreign investors basically only a 49% share in the perspective ventures discourages them from investing in the high-tech area. The second claims that due to the additional costs associated with investments in CMEA countries, Western investors must be offered more privileges to counter-balance the costs involved. The third one, on the other hand, claims that the situation observed is due to the initial period of foreign investment activities, which attracts only special types of adventurous investors looking for small investments with big profit. They will eventually disappear to be substituted by more serious investors.

It is believed that all three elements are currently at work.

As a follow up of this analysis, an assessment of the dynamics of the technological position of the CMEA countries was undertaken.

The resulting picture is extremely differentiated. It is bright as far as CMEA's scientific and technological infrastructure is concerned. It is much less positive with regards to the results of R+D activities, i.e. patenting output. The region as a whole neither managed to attract many foreign patents nor has it managed to improve its position in foreign patenting over the last two decades. The available statistics indicate that the CMEA countries enjoyed comparative advantages in old traditional fields of technology while being at disadvantage in the currently leading industries.

The technological advances of the CMEA countries in applied (i.e. actually used) technology are even less convincing. Although the countries in question managed to bridge the gap vis-à-vis the developed world over the last two decades, they lost their position with respect to newly industrialized countries. The latter managed to replace the CMEA countries in some products and markets as well as to grasp more sophisticated parts of these markets.

No trace of a dramatic impact of foreign technology inflows on regional technological build-up could be detected.

2. Some General Conclusions

What conclusions then can be drawn from the analysis performed? The first and most obvious is that technological borrowing is but only one component of the national (regional) technological build-up. To produce a positive technological multiplier, an innovation-conducive national environment and a relevant technological infrastructure must exist. The CMEA countries, while possessing extensive technological infrastructure, are largely deprived of a innovation-conducive environment. Under these circumstances, technological import basically served as a tool for overcoming current technological bottlenecks, in view of unavailable local resources. It did not, however, provide an additional impetus for local technological drive. Therefore, the results of the import operations faded out rather quickly.

An innovation-conducive national environment is merely a prerequisite for successful technological catching up via technological borrowing. Equally important are the national economic sectors to be fed with foreign technology, the proper composition of possible technology transactions, an adequate degree of industrial protection, as well as an externally-oriented foreign economic policy are only a few points to be mentioned. As CMEA's experiences indicate, the same systemic characteristics may yield different technological results due to differences in national economic and technological policies.

The benefits of technology import are heavily influenced by its characteristics per se. The principal issue is what sort of technology is being infused to the importing country. Low grade technology will produce minor results, whereas sophisticated products may have more profound effects, providing a proper absorbing capacity exists. The CMEA countries relied largely on low grade technologies, thus allowing for only limited multiplication effects.

Finally, it should be stressed that the possibilities offered by technology import are highly dependent on the conduciveness of the international transfer environment. This includes such elements as the supplying country's readiness to transfer the required technology, their goodwill in the trouble shooting stage, opening their markets for subsequent export operations and assisting in securing the necessary financing. The lack of a conducive environment for technology transfer increases the transactional costs, narrows the spectrum of available technologies and limits the size of the necessary import.

The negative experiences of the CMEA countries in this area represent an extreme contrast to the positive ones made by such countries as Japan, South Korea and Israel.

Literature

1. A. McAucley - The incompatibility of central planning and rapid innovation, mimeo, University of Essex, 1984.
2. R. Amman, J.M. Cooper (eds.) - Industrial innovation in the Soviet Union, New Haven, Yale U.P. 1977.
3. R. Amman, J. Slama - The organic chemical industry of the USSR: a case study in the measurement of comparative technological sophistication by means of kilogram-prices, Research Policy, no. 5/1976, pp. 302-326.
4. J.W. Abernathy, J.M. Utterback - Patterns of industrial innovation, Technology Review, Vol. 80, Cambridge, Mass. 1978, no. 7, pp. 40-47.
5. L. Balcerowicz - Niektóre problemy innowacji technicznych w gospodarce socjalistycznej, SGPS, Warszawa 1978, mimeo
6. T. Baumgartner - Transfer of technology: production of development or reproduction of dependency, in: D. Ernst (ed.) - The new international division of labour, technology and underdevelopment, Campus Verlag, Frankfurt/New York 1980, pp. 575-591.
7. J.S. Berliner - The innovation decision in Soviet industry, Cambridge, MIT Press, 1976.
8. L. Bialon, T. Obrebski (red.) - Postep naukowy i techniczny w rozwoju spoleczno-gospodarczym, WPW, Warszawa 1985.
9. J.C. Brada, J.M. Montias - Industrial policy in Eastern Europe; a three country comparison, Journal of Comparative Economics, Vol. 8, no. 4, 1984, pp. 377-420.
10. O. Börnsen, H.H. Glismann, J.E. Horn - Der Technologietransfer zwischen den USA und der Bundesrepublik, Tübingen, Mohr 1985.

11. M. Bernstein - The transfer of Western technology to the USSR, OECD, Paris 1985.

12. J. Brada - Industry structure and East-West technology transfer: a case study of the pharmaceutical industry, in: ACES, No. 22, spring 1980, pp. 31-59.

13. W. Brzost - Importowany postep techniczny a rozwój gospodarczy Polski, PWN, Warszawa 1979.

14. J. de Castro, T. Ganiatsos, A. Olechowski, H. Qagaya - Changes in international economic relations in the last two decades, Trade and Development. An UNCTAD Review, No. 5/1984, pp. 27-36.

15. J. Clark, Ch. Freeman, L.G. Soete - Long waves, inventions and innovations, Futures, Vol. 13, 1981, pp. 308-322.

16. J. Cieslik, J. Misala - Struktura wymiany towarowej Polski w 1971-80 wedlug intensywnosci czynników produkcji, Warszawa 1982, SGPS, mimeo.

17. R.W. Campbell - Soviet R+D statistics 1975-1982, Indiana University 1982, mimeo.

18. R. Claphan - Zum Einfluß der Wirtschaftsordnung auf den internationalen Technologietransfer, ORDO, Bd. 25, pp. 189-212.

19. Ch. Cooper - Science, technology and production in the underdeveloped countries: an introduction, The Journal of Development Studies, Vol. 9, October 1972, pp. 1-18.

20. J. Cornwall - Modern capitalism. Its growth and transformation, London 1977.

21. C.C. Coughlin - The relationship between foreign ownership and technology transfer, Journal of Comparative Economics, 7 (1983), pp. 400-414.

22. E.D.Y. Chen, Multinational corporations and technology diffusion in Hong Kong manufacturing, Applied Economics, Vol. 15, 1983, pp. 309-321.

23. A. Wass von Czege - Mechanismen zum intersystemaren Technologietransfer - ihre Klassifizierung und unterschiedliche Bewertung in Ost und West, IA u.U., Universität Hamburg, Forschungsbericht Nr. 11, Hamburg 1977.

24. C.J. Dahlman, B. Ross-Larson, L.E. Westphal - Managing technological development. Lessons from the newly industrializing countries, World Bank Staff Working Papers, no. 717, The World Bank 1985.

25. S. Derakhshani - Factors affecting success in international transfers of technology - a synthesis and a test of new con-

tingency model, The Developing Economies, Vol. 22, March 1984, No. 1, pp. 27-46.

26. J.H. Dunning - Towards a taxonomy of technology transfer and possible impacts on OECD countries, in: North-South technology transfer. The adjustments ahead, Analytical Studies, OECD, Paris 1982, pp. 8-24.

27. P.V. Elst - Capitalist technology for Soviet survival, Institute of Economic Affairs, Great Britain, 1981.

28. P. Evans - Dependent development: The alliance of multinational, state and local capital in Brazil, Princeton UP, 1979.

29. A. Emmanuel - Appropriate or underdeveloped technology? John Wiley and Sons, 1982.

30. D. Ernst (ed.) - The new international division of labour, technology and underdevelopment, Campus Verlag, Frankfurt/ New York 1980.

31. D. Ernst - Development policy: main characteristics of technological dependence and dominance and their consequences for national policies designed to strengthen technological capacities, UNIDO, ID/WG. 301/1, 30 May 1979.

32. R. Findlay - Relative backwardness, direct foreign investment and the transfer of technology: a simple dynamic model, The Quarterly Journal of Economics, February 1978, no. 1, pp. 1-16.

33. Z. Fallenbuchl - East-West technology transfer. Study of Poland 1971-1980, OECD, Paris 1983.

34. J.S. Fitch - International transfers of military technology: the political impact of US military aid to Latin America, Dept. of Political Science, University of Colorado, mimeo, undated.

35. E.E. Filippovskij - Obnovlenje promyszlennoj produkcji w kapitalisticzeskich stranach, Izd. Nauka, Moskwa 1978.

36. Ch. Freeman - The economics of industrial innovation, Penguin Books, 1974.

37. J. Galtung - Toward a new international technological order, Alternatives, 4 (1968-79), pp. 277-300.

38. A. Gerschenkron - Economic backwardness in historical perspective, Cambridge 1962.

39. H. Giersch (ed.) - Emerging technologies: consequences for economic growth, structural change, and employment, J.C.B. Mohr, Tübingen 1982.

40. S. Globerman, Technological diffusion in the Canadian tools and die industry. The Review of Economics and Statistics, Vol. 57, 1975, pp. 428-434.

41. E.M. Graham - The terms of transfer of technology to the developing nations; a survey of the major issues, in: North-South technologie transfer. The adjustments ahead. Analytical studies, Paris 1982, pp. 55-87.

42. S. Gomulka - Inventive activity, diffusion and the stages of economic growth, Aarhus University Press, 1970.

43. S. Gomulka - The incompatibility of socialism and rapid innovation, in: M.E. Schaffer (ed.) - Technology transfer and East-West relations, Croom Helm, 1985, pp. 12-31.

44. S. Gomulka, A. Nove, G.D. Holliday - East-West technology transfer. A survey of econometric and sectoral studies, OECD, Paris 1984.

45. Gaps in technology. General report, OECD, Paris 1968.

46. O. Granstrand, J. Sigurdson (eds.) - Technological and industrial policy in China and Europe, Technology and Culture, Occasional Report Series, no. 3, Research Policy Institute, Lund 1981.

47. P. Hanson - Trade and technology in Soviet-Western relations, London, Macmillan 1981.

48. P. Hanson - Western economic sanctions against the USSR: their nature and effectiveness, in: External economic relations of CMEA countries: their significance and impact in a global perspective, NATO, April 1983.

49. P. Hanson - The comparative economics of research, development and innovation: a survey, mimeo, CREES, University of Birmingham, undated.

50. P. Hanson, K. Pavitt - The comparative economics of research, development and innovation in East and West: a survey, Harwood Economic Publishers, Chur, London, Paris, New York, Melbourne, 1987.

51. J. Harders - Arzneimittelforschung und Industrieorganisation: DDR und Ungarn im Vergleich. Ökonomische Studien, Band 37, Stuttgart, New York 1985.

52. F. Hayek - The use of knowledge in society, American Economic Review, Vol. 37, no. 4, pp. 519-530.

53. Ch.T. Hill, J.M. Utterback - Technological innovation for a dynamic economy, Pergammon Press 1979.

54. H.D. Jacobsen - Die Technologiekontrollpolitik der Vereinigten Staaten und ihre Auswirkung auf die West-West Beziehungen, Europa-Archiv, Nr. 15, 1986, pp. 443-450.

55. N.D. Kondratiev - The major economic cycles, Lloyds Bank Review, no. 129, 1978, pp. 41-60.

56. J. Kornai - The economics of shortage, Amsterdam, North-Holland, 1980.

57. F. Levcik, J. Skolka, East-West technology transfer. Study of Czechoslovakia, OECD, Paris 1984.

58. F. List - Der internationale Handel, die Handelspolitik und der deutsche Zollverein, in: F. List - Das nationale System der politischen Ökonomie, Jena, 1920 (Third edition).

59. M. Lebkowski, J. Monkiewicz - Western direct investment in centrally planned economies, Journal of World Trade Law, Vol. 20, No. 6, November-December 1986, p. 627.

60. E. Mansfield - International technology transfer: forms, resource requirements and policies, American Economic Review, Papers and Proceedings, Vol. 65, 1975, pp. 372-382.

61. E. Mansfield, A. Romeo, S. Wagner - Foreign trade and US research and development, Review of Economics and Statistics, Vol. 61, 1979, pp. 49-57

62. S. Magee, Information and multinational corporation: an appropriability theory of direct foreign investment, in: J.N. Bhagwati (ed.), The new international economic order: The North/South debate, Cambridge, MIT Press 1977, pp. 317-340.

63. C.H. McMillan - The foreign investment activitiy of the Comecon countries: actors and strategies, University of Reading, Discussion Papers in International Investments and Business Studies, No. 73, June 1983.

64. C.H. McMillan - Multinationals from the second world, MacMillan Press, London 1987, pp. 38-39.

65. M. Malecki-Spolki z udzialem kapitalu zagranicznego w europejskich krajach socjalistycznych, Warszawa 1987, mimeo.

66. H. Majer - Die technologische Lücke zwischen der Bundesrepublik Deutschland und den Vereinigten Staaten von Amerika, J.C.B. Mohr, Tübingen 1973.

67. K. Marx, F. Engels - The Communist Manifesto, 1951.

68. G. Mensch - Das technologische Patt, Umschau Verlag, Frankfurt 1975.

69. G. Mensch - Zur Dynamik des technischen Fortschritts, Zeitschrift für Betriebswirtschaft, 41, 1971, pp. 295-413.

70. Measuring the changing technological level of international trade flows by means of unit values: some empirical findings, ECE, UN, Trade IR.463, 12 August 1983, p. 2.

71. J. Monkiewicz - Multinational production enterprises: a preliminary overview, UNIDO/PC.121. September 10, 1985.

72. J. Monkiewicz - Technology exports from developing countries. Dimensions, nature, potentials and issues, UNIDO/IS.525, March 28, 1985.

73. J. Monkiewicz - Miedzynarodowy transfer wiedzy techniznej. Elementy teorii i polityki, PWN, Warszawa 1981.

74. J. Monkiewicz, J. Maciejewicz - Technology export from the socialist countries, Westview Press, Boulder and London, 1986.

75. J. Monkiewicz (red.) - Polska innowacyjna, IWZZ, Warszawa 1985.

76. L.K. Mytelka - Regional development in a global economy; the multinational corporation, technology and Andean integration, New Haven, Yale University Press, 1979.

77. L.A. Niefiodow - The effects of East-West conflict on international competition for high technology, mimeo, undated.

78. K. Pavitt, L. Soete - International differences in economic growth and the international location of innovations, SPRU, Brighton, May 1981.

79. K. Pavitt, L.G. Soete - International dynamics of innovation, in: H. Giersch (ed.) - Emerging technologies: consequences for economic growth, structural change and employment, J.C.B. Mohr, Tübingen 1982.

80. K. Pavitt - Patent statistics as indicators of innovation activities: possibilities and problems, SPRU, University of Sussex, December 18, 1983, mimeo.

81. F. Perroux - L'économie du XXe siècle, Press Universitaires de France, 1961.

82. F. Perroux - The domination effect and modern economic theory, in: K.W. Rotschild (ed.) - Power in economics, Penguin 1971, pp. 56-74.

83. F. Perroux - Une théorie de l'économie dominante, Economie Appliquée, Archive de l'I.S.E.A., April-September 1948, nos. 2-3.

84. F. Perroux - Le dynamisme de la domination, Economie Appliquée, no. 2, 1950.

85. F. Perroux - La concurrence et l'éffet de domination, Banque, May 1952.

86. F. Perroux - Wirtschaft und Macht, Verlag Paul Haupt, Bern und Stuttgart, 1983.
87. G. Pollak, J. Riedel - Das Engagement deutscher Unternehmen in Entwicklungsländern - Stand und Perspektiven, Ifo-Schnelldienst 21/1984.
88. M.V. Posner - International trade and technical change, Oxford Economic Papers, Vol. 13, 1961, pp. 323-341.
89. Production and use of industrial robots, ECE, UN, New York 1985.
90. K. Poznanski - The environment for technological change in centrally planned economies, Dept. of Economics, Rensselaer Polytechnic Institute, October 1984.
91. The rate and direction of inventive activity, Princeton University Press, 1962.
92. N. Rosenberg - Perspectives on technology, Cambridge University Press, 1976.
93. N. Rosenberg - The international transfer of industrial technology: Past and present, in: North-South technology transfer. The adjustments ahead, Analytical studies, OECD, Paris 1982, pp. 25-54.
94. J. Röpke - Die Strategie der Innovation, J.C.B. Mohr, Tübingen 1977.
95. J. Röpke - Der Importierte Fortschritt. Neuerungsimport als Überlebensstrategie zentralkoordinierter Systeme, ORDO, Jahrbuch für die Ordnung von Wirtschaft und Gesellschaft, Vol. 27, pp. 223-241.
96. S. Rosenblatt (ed.) - Technology and economic development: a realistic perspective, Westview Press 1979.
97. J. Ruszkiewicz (red.) - Problemy planowania i zarzadzania w cyklu N-T-P, NOT, Warszawa 1977.
98. F. Sagasti - Technology-planning and self-reliant development: a Latin American view, N.Y., Praeger Publishers, 1979.
99. M.E. Schaffer (ed.) - Technology transfer and East-West trade, Croom Helm, London, Sydney 1985.
100. J.J. Servan-Schreiber - Die amerikanische Herausforderung, Hamburg 1968.
101. H.W. Singer, L. Reynolds - Technological backwardness and productivity growth, The Economic Journal, March 1975, pp. 873-876.
102. M. Simai - International technology transfer and economic development in the late 20th century, Trends in World Econ-

omy, No. 48, Hungarian Scientific Council for World Economy, Budapest 1984.

103. J. Sigurdson - Japan's high technology race; the information technologies, Technology and Culture, Occasional Report Series, no. 8, Research Policy Institute, Lund 1983.

104. K.E. Schenk - Märkte, Hierarchien und Wettbewerb - Elemente einer Theorie der Wirtschaftsordnung, München 1981.

105. K.E. Schenk (ed.) - Studien zur politischen Ökonomie, Ökonomische Studien, Band 32, Stuttgart, New York 1982.

106. K.E. Schenk (ed.) - Vergleichende System- und Industriestudien - Ein Institutional Choice Ansatz, Ökonomische Studien, Band 34, Stuttgart, New York 1983.

107. S.G. Schoppe - Die Sowjetische Westhandelsstruktur - ein außenhandelstheoretisches Paradoxon, Gustav Fischer Verlag, Stuttgart/New York 1981.

108. S.G. Schoppe - Technologietransfer als Strukturdeterminante des Ost-West-Handels, Schriften des Vereins für Sozialpolitik, Gesellschaft für Wirtschafts- und Sozialwissenschaften, Neue Folge, Band 116, 1981, pp. 651-664.

109. T. Smith - The underdevelopment of development literature: the case of dependency theory, World Politics, 31, January 1979, pp. 247-288.

110. J.A. Schumpeter - Capitalism, Socialism and Democracy, New York, Harper and Row 1950.

111. A. Schüller, H. Leipold, H. Hamel (eds.) - Innovationsprobleme in Ost und West, Gustav Fischer Verlag, Stuttgart/New York, 1983.

112. U.Chr. Täger - Untersuchung der Aussagefähigkeit von Patentstatistiken hinsichtlich technologischer Entwicklungen, IFO, Studien zur Industriewirtschaft, Nr. 17, München 1979.

113. C. Vaitsos - Government policies for bargaining with transnational enterprises in the acquisition of technology, in: J. Ramesh, Ch. Weiss (eds.) - Mobilizing technology for world development, N.Y. Praeger 1979.

114. R. Vernon (ed.) - Technology factor in international trade, Columbia University Press, N.Y./London 1970.

115. T. Veblen - Imperial Germany and the industrial revolution, London 1915.

116. L. Westphal, L. Kim, J.C. Dahlman - Reflections on Korea's acquisition of technological capability. The World Bank, Discussion paper, report no. 77, April 1984.

117. O. Williamson - Markets and hierarchies: analysis and antitrust implications, New York/London 1975.
118. O. Williamson - The modern corporation: origins, evolution, attributes, in: Journal of Economic Literature, Dec. 1981, pp. 1537-1568.
119. H. Wienert, J. Slater - East-West technology transfer. The trade and economic aspects, OECD, Paris 1986.
120. E. Zaleski, H. Wienert - Technology transfer between East and West, OECD, Paris 1980.

Aus unserem Programm:

Uwe Holl, Malcolm Trevor (Eds)
Just-in-Time Systems and Euro-Japanese Industrial Collaboration.
1988. Ca. 110 Seiten. ISBN 3-593-33970-6

What are the benefits of JIT production and marketing systems?
What do the Japanese expect form their suppliers? What do
Western firms have to do to get business from Japanese companies? Industrialists, purchasers, suppliers and trade unionists discuss
how to integrate JIT systems and how firms can gain an advantage in times of increasingly sharp worldwide competition.

Malcolm Trevor (Ed.)
The Internationalisation of Japanese Business
European and Japanese Perspectives.
1987. 212 Seiten. ISBN 3-593-33816-5

Japanese Companies and Strategies in the Fields of Electronics,
Financial Institutions, Automobiles and Pharmaceuticals - Nissan's
International Strategy - The Investment Strategy of Korean
Transnational Companies - Japanese Market Entry Strategy -
Innovations in Collectiv Bargaining - Communication Structures -
Transferability of Management Style - Comparative Values of
British, German and Japanese Managers Experiences and Prospects
for Euro-Japanese Collaboration - Toshiba's Approach to Purchasing
- Supplying Components to Japanese Companies - Integrating Just-
in-Time into a Total Production and Marketing System - A British
Trade Union View.

Campus Verlag - Westview Press

East Asia. International Review
of Economic, Political, and Social Development.

East Asia. Volume 1.
Japan: Problems of a Post-Industrial Society - China's integration
into the world Economy.
1983. 264 Seiten.
ISBN 3-593-33208-6. ISSN 0723-8398

East Asia. Volume 2.
Japan as a model for the First and/or Third World.
1984. 212 Seiten.
ISBN 3-593-33296-5. ISSN 0723-8398

East Asia. Volume 3.
Patterns of Outward-Centred Development - Japan and Intra- and
Interregional Cooperation.
1985. 250 Seiten.
ISBN 3-593-33511-5. ISSN 0723-8398

East Asia. Volume 4.
1987. 236 Seiten mit 12 Abbildungen und 25 Tabellen, geb.
ISBN 3-593-33773-8. ISSN 0723-8398

East Asia. Volume 5.
1989, geb. Ca. 240 Seiten.
ISBN 3-593-34036-4. ISSN 0723-8398

Campus Verlag - Westview Press